Modeling Dynamic Systems

T0189507

Series Editors
Matthias Ruth
Bruce Hannon

For further volumes:
http://www.springer.com/series/3427

James D. Westervelt · Gordon L. Cohen
Editors

Ecologist-Developed Spatially Explicit Dynamic Landscape Models

 Springer

Editors
James D. Westervelt
Construction Engineering Research
Laboratory
US Army Engineer Research
and Development Center
Champaign, IL, USA

Gordon L. Cohen
Information Technology Laboratory-IL
US Army Engineer Research
and Development Center
Champaign, IL, USA

Operational copies of the models mentioned in this book are
available at http://extras.springer.com/2012/978-1-4614-1256-4.

ISBN 978-1-4899-8779-2   ISBN 978-1-4614-1257-1 (eBook)
DOI 10.1007/978-1-4614-1257-1
Springer New York Dordrecht Heidelberg London

Springer is part of Springer Science+Business Media (www.springer.com)

Editors' Foreword

We have given ourselves the job of helping to persuade you—a creative ecologist or social scientist—that you have all the necessary capabilities to begin capturing your unique expertise in simple, powerful simulation models that codify your knowledge into a computerized analytical tool. Your model gives you the opportunity to share your individual insights with your community of peers in the form of an easy-to-use, science-driven computer program that they can in turn examine, use, extend, and repurpose for their own work.

Simulation modeling is no longer the exclusive domain of elite computer scientists and programmers. Practical and expedient models now can be written without any mastery of low-level computer languages, numerical methods, or interface design. Simulation modeling platforms are now available that facilitate experimentation without bogging down the model builder in complicated software compiling tasks or graphical output issues. Powerful, user-friendly model-development tools have emerged—both open source programs and commercial packages—that can be mastered by anyone who has expert knowledge of a system, a fundamental understanding of desktop computers, and willingness to learn how to use software that is considerably less complicated than the everyday "office" applications that vex us all from time to time. You will find simulation modeling to be a gratifying and highly empowering skill if you are interested in:

- Harnessing computer power to reflect the implications of your intuitive understanding of a system, and make supportable predictions based on them.
- Verifying whether your understanding of a system can be codified in a way that replicates known system behaviors.
- Personally transcribing your intuitive expert knowledge into a transparent, science-based framework without asking computer programmers to intervene.

In the preface to this book, Dr. Bruce Hannon describes how he has encouraged a generation of social science and ecology students to climb the modest learning curve within a few class sessions, and then apply their skills to building operational simulation models in workgroups of two to eight. In his classes and the preface, Dr. Hannon emphasizes the benefits that students will gain by acquiring formal, but expedient, simulation modeling skills. He illustrates how an individual's deep understanding of a system's dynamics and behavior can readily be captured in a form that computers can process to unveil hidden implications of system processes that would otherwise probably evade conscious thought. Modeling enables you to do what you are good at—describing the system—while enlisting the computer to make supportable projections based on your expert knowledge.

This book is divided into two parts: (1) a technical orientation for prospective modelers and (2) examples of expedient operational models developed using the methods and tools described in Part I. The first part is intended especially for readers with no substantive experience in model building, but it includes insights that should benefit all modelers.

Chapter 1 addresses the topic of "modeling reluctance," for lack of a better term, that often inhibits ecologists and social scientists from acquiring model-building capabilities. As Dr. Hannon notes in his Preface, this inhibition can affect researchers like you, who have built a large store of technical expertise based on both direct observations from the field and an intuitive capability for drawing accurate inferences about future system behavior based on changes to the environment. If computer programming and higher mathematics are far removed from your daily practice, it is not surprising that you would be skeptical about how these disciplines might contribute to your work. Chapter 1 makes it clear that model building does not require high levels of computer or mathematical expertise and explains that modeling is already part of your everyday cognitive processes.

Chapter 2 describes a general process by which multidisciplinary groups may use relatively simple software tools to model relatively complex domains. It provides a general project roadmap to help multiple researchers from different disciplines work efficiently and harmoniously toward creating a rich simulation model in a very reasonable amount of time. These working guidelines have been used successfully at the University of Illinois for more than a decade to teach nonprogrammers how to develop dynamic simulation models working in a computer lab environment for several hours a week over a single semester. Most of the models presented in the second half of the book were created as class projects by multidisciplinary groups ranging from two to eight in size. Many of the team members were new to computer-based modeling.

Chapter 3 introduces you to NetLogo (Wilensky 1999), the model-development environment that was used to construct the models documented in the second part of the book. NetLogo is a free, public domain model-building software platform that enables you to describe the behavior of individuals within the spatial environment they inhabit. The individuals can interact with each other and their environment, and the environment itself may change according to its own dynamics. The chapter also provides grist for traditional computer programmers: a short introduction to Repast Simphony, a free, open source agent-based modeling package developed by Argonne National Laboratory, U.S. Department of Energy (http://repast.sourceforge.net/). Repast offers a migration path from simple NetLogo models to more challenging simulation modeling environments preferred by computer scientists. The fall 2010 release of Repast includes the ReLogo framework, which converts NetLogo models to Repast compatibility. Once converted, a computer programmer can then integrate the NetLogo model with other models, run the model on more powerful machines, and visualize and analyze model outputs in many useful ways not natively available in NetLogo.

Part II presents 11 simulation models, as documented using the Overview, Design concepts, and Details (ODD) protocol (Grimm et al. 2006). This protocol, developed cooperatively by 28 professional modelers, is a standardized model documentation specification intended to help a model builder clearly communicate the essential contents of a simulation model as well as its assumptions and scope. The purpose of the ODD protocol is to make the contents of a simulation model transparent to a reader who has some knowledge of the specific technical domain for which the model is built. This content format also makes it easy for the reader to evaluate similar models side by side.

All 11 models documented in Part II (Chaps. 4–14), and their data sets, are available for download and use.[1] These models were developed in NetLogo (Wilensky 1999), and your learning experience will be greatly enhanced if you load and run each model on your computer as you are reading about it in the book. You may download a full, operable version of NetLogo from http://ccl.northwestern.edu/netlogo/. All models presented in this book have been tested to run in NetLogo 5.0. Because NetLogo is programmed in Java (Oracle, Redwood Shores, CA), it operates on computers running Microsoft Windows, Macintosh OS X, or Linux.

Most of the models presented in Part II were developed and authored by students who took a University of Illinois spatial simulation-modeling course taught by Dr. Hannon, Dr. Charles Ehlschlaeger, and Dr. Jim Westervelt. They are grouped as individual-based models (IBMs) representing animal populations in the wild (Chaps. 4–8), a river nutrient model (Chap. 9), patch and inter-patch valuation models (Chaps. 10–12), and social models (Chaps. 13 and 14).

The first two models, fire ants (Chap. 4) and newts (Chap. 5), were developed by students in the class to explore, respectively, control measures for red imported fire ants (RIFA) in Texas and forecasting responses of striped newts to rainfall patterns in Georgia. In both cases, a pair of students new to simulation modeling turned literature reviews and interviews with experts into conceptual and then working models. The next two chapters consider the gopher tortoise, a species at risk, in the southeast United States. Chapter 6 captures a research effort that did not include direct involvement by an ecologist familiar with the gopher tortoise, but did involve experienced modelers. Conversely, the model in Chap. 7 was created by a team of ecologists familiar with the tortoise but without any experience in simulation modeling. This team quickly achieved proficiency with NetLogo.

The feral hog model described in Chap. 8 was developed by a team of seven graduate students, none of whom had previously written software. Their purpose was to test the hypothesis that adding a contraceptive program to an existing hunting policy would improve the control of wild swine on a military installation in Georgia. At one point during the course, the sound of virtual gunshots cracked out through the lab from computer speakers—NetLogo-generated hunters applying

---

[1] Operational copies of the models are available through http://extras.springer.com.

"control measures" during a demonstration of work in progress. The model proved so useful for testing the advantages of a proposed contraceptive program that one student, now the lead author of Chap. 8, was funded to further develop the model.

Chapter 9 explores nutrient cycling in the Mississippi River, taking into account the movement of nutrients via water currents in a pool on the river. River nutrients cycle through several trophic layers as the water flow moves components of the system downstream. This effort began with a nonspatial model written a decade earlier that was adapted to produce spatial output in NetLogo.

The next three chapters explore the value of land in terms of its contribution to the viability of a population. Habitat patches are analyzed in Chaps. 10 and 11. The first of those traces the lineage of populations in patches over time with respect to the original home of the original ancestors to reveal the relative connectivity among all pairs of patches. The second documents a model developed to reveal the relative value of each patch supporting a metapopulation in terms of sustaining the viability of the metapopulation. The intent of this second model is to support the development of an equation into which certain characteristics of patches, easily measured in the field, could be used to compute a "patch valuation" estimate. Chapter 12 looks at the value of land between patches for supporting inter-patch migration, which is necessary to connect populations into a broader metapopulation. This project translated a published supercomputer-based model into the NetLogo modeling system. The result is a very accessible model useful for experimentation and potential extension.

The final two chapters explore social science models that extend beyond natural ecosystems. Chapter 13 considers a model that forecasts urban residential growth patterns within a county based on the relative attractiveness of land to that growth. Domestic violence is the subject of the model documented in Chap. 14. The help-seeking behavior of violence victims is explored in a way that makes it possible to test policy impacts on violence rates.

Each of these models demonstrates how students and researchers have captured their understanding of dynamic spatial systems using a simulation modeling software package. The models make it possible for users to experimentally manipulate the system to test its response when subjected to alternate assumptions, conditions, or scenarios. Our hope is that these examples will encourage you to do the same!

Champaign, IL, USA   James D. Westervelt
                                                                         Gordon L. Cohen

References

Grimm V, Berger U, Bastiansen F, Eliassen S, Ginot V, Giske J, Goss-Custard J, Grand T, Heinz SK, Huse G, Huth A, Jepsen JU, Jørgensen C, Mooij WM, Müller B, Pe'er G, Piou C, Railsback SF, Robbins AM, Robbins MM, Rossmanith E, Rüger N, Strand E, Souissi S, Stillman RA, Vabø R, Visser U, DeAngelis DL (2006) A standard protocol for describing individual-based and agent-based models. Ecol Model 198(1–2):115–126
Wilensky U (1999) NetLogo: computer software. Center for Connected Learning and Computer-Based Modeling, Northwestern University, Evanston. http://ccl.northwestern.edu/netlogo/

Preface

The Simulation Model: A Left-Brain Tool for Right-Brain Scientists

In the domain of ecology, there exists a huge source of information that is largely undocumented and therefore unavailable to practitioners. It is expertise that is sequestered in the individual minds of many field ecologists and rarely captured in a form that is readily accessible by the greater community of practice. The nature of this expertise differs depending on the interests and working style of the practitioner. Some ecologists seek documentable precision in knowledge by investigating natural systems through the collection of large data samples capable of producing statistically verifiable insights. This quantitative approach can offer intimate and accurate understandings of small subsets of an ecosystem. Other ecologists develop their knowledge by conducting diverse case studies designed to inform a larger overview. Both approaches lead ecologists to develop valuable insights on how ecosystem components function and interact. Each individual's growing expertise constitutes a part of a rich, but uncompiled, knowledge base. It is available to the possessor and associates for specific projects or applications, but it remains generally, if unintentionally, concealed from the greater community of practice.

Psychology informs us that people have two different modes of thinking, each of which roughly correlate to one brain hemisphere or the other. Right-brain thinking is considered to be more creative, intuitive, holistic, and spontaneous, while left-brain thinking is considered to be more methodical, logical, linear, and analytical. In terms of ecological research, the synthesis of big-picture results from many case studies represents a right-brain approach, and the development and analysis of large data samples represents a left-brain approach. But because there is little overlap in the two approaches, we often have to choose between keen but unverifiable intuition, on one hand, and hard but never-complete data on the other. And these differences pose an understanding gap between experts from the two different methodological approaches.

This gap may be illustrated by the following scenario. Over many years, a field ecologist develops deep, intuitive insight into an ecosystem that makes it possible for him or her to forecast the consequences of proposed management actions on an ecosystem, often with a very high level of confidence. A planner who is considering new management initiatives may seek out the insights of the seasoned expert, whose reputation the planner knows and trusts. The field ecologist's expertise is often rooted more deeply in experience and intuition than in peer-reviewed research. If he or she wants to disseminate those insights to others beyond the immediate research team or work group, prospective users must be able to verify the validity and applicability of that expertise.

One approach the field ecologist can take to disseminate the use of hard-won technical insights is to apply left-brain skills to what is already understood intuitively—to explicitly identify and analyze the cause–effect relationships that lie beneath the intuitive knowledge. A computer simulation model is an excellent tool for capturing and representing such technical knowledge in a way that is highly explanatory and well documented. A simulation model can employ validated algorithms plus data and alternate assumptions to reflect the field ecologist's insight into the implications of environmental change or management actions. Simulation results can be compared with the ecologist's "instincts," both to assess the validity of the model and to further illuminate the right-brain thinking behind it. Any gaps revealed between simulation results and the ecologist's deep understanding can be considered and addressed. As the model is refined and simulation results match the right-brain understanding of the system, the ecologist achieves an analytical validation of ideas that may previously have been beyond the reach of the left brain. At that point, the model is ready to share and to apply to specific cases, which can help decision makers and the general public develop improved impact analyses and policy alternatives.

For more than 25 years, I have taught life science students at the University of Illinois at Urbana-Champaign how to simulate dynamic biological phenomena on computers. It is my favorite activity as a professor. Over the years, students have modeled a full gamut of biological activity, ranging from the disciplines of microbiology to genetic engineering, and covering the dynamics of the individual cell, bacteria, individual plants or animals, and large collections of organisms. I try to help them learn that the intent of building these models is to better understand function and limits for the ultimate purpose of informing good management practice.

Regardless of my enthusiasm and best efforts, I have not had unqualified success at teaching my students why I believe that dynamic modeling and the acquisition of systems thinking capabilities are so essential to their future work. Below, I explore why this has happened and what might be done about it. I also will clearly lay out the general benefits of modeling. Students do well in my course in part because it is tailored to minimize reliance on sophisticated mathematics and programming. We are fortunate that model-building computer environments such as STELLA (isee systems, Lebanon, NH) and NetLogo (Wilensky 1999) are now available to help students to quickly and easily capture and document their ideas about biological dynamics as computer simulation models. The models created using these tools enable my students to clearly explain to me, to their other professors, and to the professional community the structure and dynamics of those areas where their specific interests lie.

Although students are not actively discouraged from model building by their thesis supervisors, they are not actively encouraged to investigate it, either. Most of my students have enrolled in my modeling courses more on their own volition than on someone's advice. A second inhibitory factor is that modeling must be practiced continuously in order to develop skills and internalize them. Because the typical students in master's or doctoral programs in this area are under high demand to perform laboratory and field experiments, they find little time or incentive to build models for the purpose of capturing their understandings of how systems work.

Ecology-oriented students are traditionally focused on hypothesis-driven case studies and huge data collection projects that allow them to draw statistical inferences about how their systems function. This approach to research rarely allows one to infer behavior at one level from behavior observed at a lower level, or at one location in a landscape behavior observed in another. It does not help students to formalize first-principle understanding of the cause-and-effect functioning of their systems.

I have often speculated why this is the case. Is it because they are not trained in simulation modeling at an earlier age, as are engineering students, for example? Or do these students imagine that modeling and simulation require skills that are beyond their reach? (The ease of modeling using new and evolving software environments could dispel that notion, given some introductory hands-on instruction.) Or do such students really so love nature that they simply seek the means to dwell within it through ecological fieldwork? I would argue that this love of nature might be significantly enriched by starting the journey with a set of hypotheses, followed by a modeling exercise that can verify and improve their understanding of their system. A model offers students a means for testing their assumptions and questions and for identifying the parameters that must be investigated and verified by lab or fieldwork. It also can help students understand which parameters are the most important and which can be reliably derived from the literature.

I begin each course by sharing the idea that education has been evolving since literacy was solely found within the monastery, through the time when we realized that numeracy was required to distinguish the importance of our assumptions, to the present, when we find it necessary to add systems thinking to the list. Systems thinking helps us to more accurately formulate pertinent questions about the phenomena that interest us. As with the acquisition of literacy and numeracy, skill in systems thinking improves with practice. And the level of practice increases with improved understanding of the power of systems thinking. This explanation leads to my discussion of the power of systems thinking, and how dynamic systems simulation on the computer provides the key to this power.

It helps us to understand that we all model the dynamics of the world around us. We instinctively know how to duck a stone thrown at us, we know how to safely cross a street in fast, heavy traffic, and how to hit a baseball. We do this by first formulating a mental model of the process and the probable consequences of various alternative actions. We evolve this model by our own trial and error and by observation of the actions of others. Given that we all routinely construct mental models, it should come as no surprise that we can increase the complexity and explanatory power of those models by extending them with computer power. The application of computers to our models of the world expands the reach of our mind in a similar way that the telescope and microscope extend the reach of our eyes.

When we try to extend our mental models exclusively through thought to solve complex social, political, or economic problems, for example, we encounter three specific difficulties. First is the uncertainty of our grasp of the important features of the problem; second is the effects of responses to our interventions or to internal forces driven by complex feedback loops; and third is the delay between the interventions (or forces) and the reactions to them. This uncertainty—these feedbacks and delays—can so complicate the dynamics of a system that the human minds cannot account for them all unaided. Society has reached the point where the complexity of environmental, interpersonal, and interagency connections is growing faster than the human mind can evolve to comprehend them. So instead of waiting for evolution, humans invent the means to extend our senses—and now, our capacity to apply logic—in order to master the complexities of the system in a timely way. To my mind, that is the great promise of simulation modeling technology.

But what are the specific benefits of computer-aided dynamic modeling? Over the years, with the help of many others, I have compiled a list of such benefits. Presented roughly in order of importance, dynamic modeling:

1. Can highlight the gaps in our understanding of the system processes. The construction of a computer model requires us to systematically lay out the stocks and flows within a system and to identify the nature of the systems controls. It helps us to establish a hierarchy of importance of system parameters. It enables us to identify and challenge the assumptions behind our understanding of the process. Simulation results, along with clear documentation of the model structure, make it possible to provide a common frame of reference for all those involved in studying and managing the system.
2. Provides a system memory. Model building is the process of formally building and joining models of the component parts of a system to create a published description of it. Every validated model iteration contributes to a more realistic model of the whole system for everyone who is interested.
3. Reveals "normal" system performance. Large changes in a system's behavior are, many times, just rare events that a good system model would show to be expected and at what frequency. Managers of such systems, without the aid of a model, tend to implement changes based on the occurrence of these rare but potentially expectable events. Such management actions, if based on a misdiagnosis of the environmental stress, can produce delayed reactions that have the potential to throw the system into disarray.
4. Allows testing *what-if* scenarios and experimentation with various kinds and levels of system management. A dynamic model makes it possible to see what happens when a system fails without any real-world consequence, and at far lower cost than witnessing an actual failure of the real system.
5. Provides quantitative information about the system operation at organizational levels (e.g., landscape or biome) and time scales (e.g., centuries) not ever experienced by observers of the real system.
6. Reveals emergent properties of the system, such as reactions and new states anticipated by no one involved in the study of the system. In other words, a dynamic model makes it possible to develop realistic predictions of a complex system under dynamic conditions.

7. Allows for "mediated modeling," which involves all appropriate experts and stakeholders, and facilitates the development of consensus in complex or controversial situations. Current software is user-friendly and transparent enough that novices can quickly understand that their views are being accurately captured in the model. Once this is accomplished for all of those involved, the simulation results are more credible and, therefore, more readily accepted by all. Mediated modeling also can shed light on the accuracy of contending theories about system functions.
8. Promotes the accurate formulation of novel, previously unanticipated questions about system performance.

If these benefits provide sufficient motivation for the student to begin the investigation and practice of model building, then it is appropriate to generally outline what is involved in the modeling process.

The most suitable environment for creating spatially explicit dynamic models will be simple to learn but capable of handling high complexity. It should serve as a stepping stone to compiled modeling languages such as C+ when the form of the model has become fixed and intensive parameter testing is required. The programming language should make maximum use of symbols for the state and control variables in order to take advantage of our ability to quickly understand such symbolism. The programming language itself should be capable of handling statements in English-like language and provide efficient input from data sources. The language should be capable of graphical data input and have some ability to model spatially. It should allow easy testing of the effects of parameter variation.

STELLA (http://www.iseesystems.com/) fully meets these requirements, so it is ideal for those who are beginning to model and wish to explore while easily changing model structure and controls. STELLA is a simulation modeling environment that allows one to graphically capture the cause–effect relationships of a system that affect state variables. Equations and logic are then added to determine rates of flows in the state variables during a predetermined time step. When the model is finished to the developer's satisfaction and is ready for extensive parameter sensitivity testing, curve fitting the model results to known data, or optimizing a certain state variable, another program is needed. My students use Berkeley-Madonna (http://www.berkeleymadonna.com/) to transform a STELLA model into a compiled form that runs many times faster than it can natively in STELLA. The Berkeley-Madonna program (1) runs extensive parameter sensitivity trials, (2) fits the model results to a given set of data, and (3) optimizes a given state in the model. The second item treats the model as though it was a regression "equation," allowing that equation to embody all of our specific understanding of the system.

STELLA is most useful for modeling systems that are homogeneous in space. If the dynamic system model requires specific location-dependent detail, one can develop the model for each cellular space (or cell) in STELLA, and then translate those into the NetLogo modeling environment (Wilensky 1999, http://ccl.northwestern.edu/netlogo/) to capture the spatial dynamic process. Each parameter is set using a digital map to represent its geographical variation.

The NetLogo environment is the best compromise between the simple programming requirements of STELLA (which is ideal for either a single-cell model or a spatial model with no more than, say, 25 cells) and the complex programming required to knit thousands of cellular models together into a dynamic whole. One can learn a significant amount from a STELLA model, and it is always useful to begin one's ecological modeling there. But the resulting model will need to be restated in NetLogo with added programming to incorporate the maps of the constants and initial state values. It is quite possible that the slightly more demanding programming skills needed for using NetLogo will eventually evolve into an even simpler procedure. Our practice is to use either free or commercially available software and concentrate on the process of modeling instead of developing a spatial modeling program of our own.

Having taught the spatial dynamic modeling course at the University of Illinois for more than 20 years, with the help of Dr. James Westervelt and Dr. Charles Ehlschlaeger, we have evolved what I believe to be the best current way to learn the process. We start the class by dividing the students into teams of two or three, with each team focusing on a specific set of modeling questions. The first 2 weeks are spent learning NetLogo, and the rest of the course is devoted to finishing the model, preparing the map data, and answering the modeling questions.

Some class projects have blossomed into large follow-on projects, including master's theses and doctoral dissertations. The LEAM urban development model (http://www.leam.uiuc.edu/) originated in this class and is now the basis of a company and a university laboratory. Our model of the Mississippi River aquatic ecosystem is another such project, having begun in the class and now the basis for a major interuniversity project. As these models matured and grew to the point of tens of millions of cells, the programs were rewritten in C++, which greatly accelerated simulation speed but required more esoteric knowledge to revise the model.

I cannot overstate to life science and social science students the importance of first formulating clear and concise questions about the phenomenon of interest. After that, they should construct a model—first in STELLA—of the part of the ecosystem that is most directly relevant to answering their questions about it. This two-phase process, if well executed, will reveal after relatively little time and expense the parameters to which the model is very sensitive. Discovering the values of these key parameters becomes the objective of their lab and field experiments. Data from the literature may be sufficient to obtain the rest of the parameters. This process reduces the overall research work and makes its progress more predictable.

Urbana, IL, USA   Bruce Hannon

Reference

Wilensky U (1999) NetLogo: computer software. Center for Connected Learning and Computer-Based Modeling, Northwestern University, Evanston. http://ccl.northwestern.edu/netlogo/

Contents

1   Never Fear: You Already Model!... 1
    James D. Westervelt and Gordon L. Cohen

2   A Collaborative Process for Multidisciplinary Group
    Modeling Projects 7
    James D. Westervelt and Bruce Hannon

3   An Introduction to the NetLogo Modeling Environment.................. 27
    David Stigberg

4   A Simulation Model of Fire Ant Competition
    with Cave Crickets at Fort Hood, Texas 43
    Bart Rossmann, Tim Peterson, and John Drake

5   Spatially Explicit Agent-Based Model of Striped
    Newt Metapopulation Dynamics Under Precipitation
    and Forest Cover Scenarios... 63
    Jennifer L. Burton, Ewan Robinson, and Sheng Ye

6   Forecasting Gopher Tortoise (*Gopherus polyphemus*)
    Distribution and Long-Term Viability
    at Fort Benning, Georgia.. 85
    James D. Westervelt and Bruce MacAllister

7   Using Demographic Sensitivity Testing to Guide
    Management of Gopher Tortoises at Fort Stewart,
    Georgia: A Comparison of Individual-Based Modeling
    and Population Viability Analysis Approaches 109
    Tracey D. Tuberville, Kimberly M. Andrews, James D. Westervelt,
    Harold E. Balbach, John Macey, and Larry Carlile

8 A Model for Evaluating Hunting and Contraception as Feral Hog Population Control Methods 133
Jennifer L. Burton, Marina Drigo, Ying Li, Ariane Peralta, Johanna Salzer, Kranthi Varala, Bruce Hannon, and James D. Westervelt

9 Spatially Explicit Modeling of Productivity in Pool 5 of the Mississippi River 151
Katherine R. Amato, Benjamin Martin, Aloah Pope, Charles Theiling, Kevin Landwehr, Jon Petersen, Brian Ickes, Jeffrey Houser, Yao Yin, Bruce Hannon, and Richard Sparks

10 Simulating Gopher Tortoise Populations in Fragmented Landscapes: An Application of the FRAGGLE Model 171
Todd BenDor, James D. Westervelt, J.P. Aurambout, and William Meyer

11 An Individual-Based Model for Metapopulations on Patchy Landscapes-Genetics and Demography (IMPL-GD) 197
Jennifer L. Burton, Richard F. Lance, James D. Westervelt, and Paul L. Leberg

12 An Implementation of the Pathway Analysis Through Habitat (PATH) Algorithm Using NetLogo 211
William W. Hargrove and James D. Westervelt

13 A Technique for Rapidly Forecasting Regional Urban Growth 223
Todd BenDor and James D. Westervelt

14 Modeling Intimate Partner Violence and Support Systems 235
Marina Drigo, Charles R. Ehlschlaeger, and Elizabeth L. Sweet

Index 255

Contributors

Katherine R. Amato Program in Ecology, Evolution, and Conservation Biology, University of Illinois, Urbana, IL, USA
amato1@illinois.edu

Kimberly M. Andrews Savannah River Ecology Lab, University of Georgia, Aiken, SC, USA
Andrews@srel.edu

J.P. Aurambout Australian Department of Primary Industries, DPI Parkville Centre, Carlton, Australia
jeanphilippe.aurambout@dpi.vic.gov.au

Harold E. Balbach Construction Engineering Research Laboratory, U.S. Army Engineer Research and Development Center, Champaign, IL, USA
Hal.E.Balbach@usace.army.mil

Todd BenDor Department of City and Regional Planning, University of North Carolina at Chapel Hill, Chapel Hill, NC, USA
bendor@unc.edu

Jennifer L. Burton Department of Natural Resources and Environmental Sciences, University of Illinois, Urbana, IL, USA
jlburton@illinois.edu

Larry Carlile Fort Stewart Army Installation, Fish and Wildlife Branch, Fort Stewart, GA, USA

Gordon L. Cohen Information Technology Laboratory-IL, US Army Engineer Research and Development Center, Champaign, IL, USA
gordon.l.cohen@usace.army.mil

John Drake School of Integrative Biology, University of Illinois, Urbana, IL, USA
jedrake@bu.edu

Marina Drigo Department of Urban and Regional Planning, University of Illinois, Champaign, IL, USA
marina.v.drigo@gmail.com

Charles R. Ehlschlaeger U.S. Army Engineer Research and Development Center, Construction Engineering Research Laboratory, Champaign, IL, USA
Charles.R.Ehlschlaeger@usace.army.mil

Bruce Hannon Department of Geography, University of Illinois, Urbana, IL, USA
bhannon@illinois.edu

William W. Hargrove USDA Forest Service Southern Research Station, Eastern Environmental Threat Assessment Center, Asheville, NC, USA
whargrove@fs.fed.us; hnw@geobabble.org

Jeffrey Houser Upper Midwest Environmental Sciences Center, U.S. Geological Survey, La Crosse, WI, USA
jhouser@usgs.gov

Brian Ickes Upper Midwest Environmental Sciences Center, U.S. Geological Survey, La Crosse, WI, USA
bickes@usgs.gov

Richard F. Lance Environmental Laboratory, U.S. Army Engineer Research and Development Center, Vicksburg, MS, USA
Richard.F.Lance@usace.army.mil

Kevin Landwehr U.S. Army Corps of Engineers, Rock Island, IL, USA
Kevin.J.Landwehr@usace.army.mil

Paul L. Leberg Department of Biology, University of Louisiana, Lafayette, LA, USA
leberg@louisiana.edu

Ying Li Department of Crop Sciences, University of Illinois, Urbana, IL, USA
yingli3@illinois.edu

Bruce MacAllister Construction Engineering Research Laboratory, U.S. Army Engineer Research and Development Center, Champaign, IL, USA
Bruce.A.MacAllister@usace.army.mil

John Macey Fort Stewart Army Installation, Fish and Wildlife Branch, Fort Stewart, GA, USA

Benjamin Martin Department of Natural Resources and Environmental Sciences, University of Illinois, Urbana, IL, USA

William Meyer U.S. Army Engineer Research and Development Center, Construction Engineering Research Laboratory, Champaign, IL, USA
William.D.Meyer@usace.army.mil

Ariane Peralta Program in Ecology, Evolution, and Conservation Biology, University of Illinois, Urbana, IL, USA
alperalt@illinois.edu

Jon Petersen U.S. Army Corps of Engineers, Rock Island, IL, USA
Jonathan.W.Peterson@usace.army.mil

Tim Peterson Department of Kinesiology, Kinesiology and Community Health, University of Illinois, Urbana, IL, USA
tim.taiji@gmail.com

Aloah Pope Department of Natural Resources and Environmental Sciences, University of Illinois, Urbana, IL, USA
pope7@illinois.edu

Ewan Robinson Department of Geography, University of Illinois, Urbana, IL, USA
ewan.robinson@gmail.com

Bart Rossmann Applied Technology for Learning in the Arts and Sciences, University of Illinois, Urbana, IL, USA
bartmann@illinois.edu

Johanna Salzer Department of Pathobiology, University of Illinois, Urbana, IL, USA

Richard Sparks Illinois Natural History Survey, University of Illinois, Elsah, IL, USA
rsparks@illinois.edu

David Stigberg U.S. Army Engineer Research and Development Center, Construction Engineering Research Laboratory, Champaign, IL, USA
David.K.Stigberg@usace.army.mil

Elizabeth L. Sweet Department of Geography and Urban Studies, Temple University, Philadelphia, PA, USA
bsweet@temple.edu

Charles Theiling U.S. Army Corps of Engineers, Rock Island, IL, USA
Charles.H.Theiling@usace.army.mil

Tracey D. Tuberville Savannah River Ecology Lab, University of Georgia, Aiken, SC, USA
tubervil@uga.edu

Kranthi Varala Department of Crop Sciences, University of Illinois, Urbana, IL, USA
kvarala2@illinois.edu

James D. Westervelt Construction Engineering Research Laboratory, US Army Engineer Research and Development Center, Champaign, IL, USA
james.d.westervelt@usace.army.mil

Sheng Ye Department of Geography, University of Illinois, Urbana, IL, USA
sheng.y.ye@gmail.com

Yao Yin Upper Midwest Environmental Sciences Center, U.S. Geological Survey, La Crosse, WI, USA
yyin@usgs.gov

Chapter 1
Never Fear: You Already Model!

James D. Westervelt and Gordon L. Cohen

The task of writing simulation models to support environmental management decision-making has historically been assigned to software development specialists working in cooperation with a technical subject matter expert. The objective of model development was, and is, to capture the expert's knowledge of a system to provide a more formal description and analysis of the system. Because few ecological management professionals have the computer science training to direct the actual model-building effort, computer specialists have carried out most of that work. Common programming languages have included FORTRAN, C, Java, and Perl. Models based on statistical analysis have been developed using scripting languages with software packages such as R, SPSS, and SAS. Spatially explicit models incorporate geographical information systems (GIS) that provide scripting languages for executing map analysis. For purely mathematical models, there are programming tools included in industry-standard packages such as Mathematica and Matlab. Because of the need to recruit computer specialists for the bulk of model-building work on behalf of the subject matter expert, model development efforts have historically been costly and time-consuming. Ecological models considered to be the most useful over time have been given names, then reused, and maintained over many years. Not surprisingly, these successes tend to produce somewhat generic results.

Ecologists who have no formal training in model development—that is, most of them—have tended to consider the computer-based simulation model as a costly "black box" tool whose utility is marred by uncertainty about how it works inside. Consequently, most professional ecologists have chosen to ignore formal modeling

J.D. Westervelt (✉)
Construction Engineering Research Laboratory, US Army Engineer Research
and Development Center, Champaign, IL, USA
e-mail: james.d.westervelt@usace.army.mil

G.L. Cohen
Information Technology Laboratory-IL, US Army Engineer Research
and Development Center, Champaign, IL, USA
e-mail: gordon.l.cohen@usace.army.mil

J.D. Westervelt and G.L. Cohen (eds.), *Ecologist-Developed Spatially Explicit Dynamic Landscape Models*, Modeling Dynamic Systems, DOI 10.1007/978-1-4614-1257-1_1, © Springer Science+Business Media, LLC 2012 of their management domains and make decisions based on their own scientific expertise, experience, and intuition. We might say that ecologists are typically most comfortable using the creative part of their brain, their right brain. Thinking with this part of the brain opens the conscious to the unregulated wanderings and musings of the subconscious where experiences are folded together to provide us with deep understandings of systems. This results in our ability to intuitively forecast consequences of actions on or changes to the systems with which we are familiar. Through this type of intelligence, instinct, and good luck, most ecologists can succeed, excel, and greatly prosper at their work. For a scientist, however, using only this approach has its shortcomings. One significant deficiency is that the individual's professional knowledge itself emerges from a kind of black box—a complex but subjective thought process in which the system is captured only informally, as far as the rest of the world knows, possibly using unquantifiable assumptions or undocumented intuitions. This is where left brain thinking, involving logic, classification, scheduling, process, and procedure, becomes valuable. By carefully looking at and describing the parts of a system, it becomes possible to build a more comprehensive understanding of the cause-and-effect relationships underlying the behavior of the whole system. When we are able to capture the assumptions and dynamics of a system formally and clearly, they become available for review by anyone who can read the language in which the knowledge is encoded. That encoding has traditionally been accomplished in print, using the language of the discipline. The same information also can be captured in computer languages, and this is highly useful because computers are very good at executing their instructions to reflect the behavior of a captured system when the state of the model is altered. When used together, captured right- and left-brain thinking can provide much more complete and compelling insights than using either exclusively.

Although the practice of modeling may seem abstruse to the reader who is not a computer programmer, we will venture to point out that you are already an expert modeler! Humans, as well as all higher animals, must reliably model the world around them in order to survive. Grazers, for example, must retain a mental image of the seasonally changing locations of food, shelter, and water. Predators and prey must develop cognitive models of the "battleground" where hunting and foraging take place. Each must be able to sense the probability of success or failure in satisfying the fundamental requirements for survival and procreation. The deer, on one hand, has a model of how close it can allow the wolf to approach in the current terrain and still be able to flee to safety; on the other hand, the wolf has a model of how to use the same terrain, wind direction, and other factors to approach the deer with stealth in order to overtake it. A tourist visiting in New York City must develop an "on the fly" model for how to efficiently cross a busy street without being run over by a yellow taxi.

Consider the components that comprise a "geospatially explicit model" of a baseball outfielder as a towering fly ball soars toward the wall in left-center. The player must rapidly calculate trajectory and peak altitude in order to determine whether to field the ball on the fly or a bounce, knowing both the static location and height of the wall and also the dynamic location of another outfielder who may be running on a converging path. Temperature affects air density and, thus, the drag of friction; and the cloud cover may or may not provide additional visual information. All these data streams feed an ad hoc model that the fielder uses to coordinate running across a landscape and reaching into space in the hope of putting out a batter, and perhaps instantly hurling the ball to an infielder to keep a runner from advancing.

Or consider a more impressive modeling project: a baby learning to walk. Lacking even the basic tools represented by a significant spoken vocabulary, the infant builds rudimentary models of his or her body in space and time, refined and combined over a year on the planet, to fully repurpose a body that was seemingly designed for a prone, static existence. This new mind, unimpeded by ideas of what may not be possible, exercises a growing spatial awareness and flaccid neck muscles to balance an outsized head upright. Within a few months, the baby can balance the whole upper body in a seated position—no hands! And then, through imitation, trial, error, continuous observation, and revision and combination of "sub-models," uses those free hands and higher-resolution spatial models to grasp objects and parental appendages in the environment. Legs learn to perambulate in response to forward guidance by a parent; lower body strength grows; sub-models of vertical balance are revised as the baby learns how to stand stably with a higher center of gravity than just a few weeks before. Somehow, using observational and experimental methods that cannot be adequately communicated in chapters of technical writing, the infant applies spatial and nonverbal conceptual models to learn the exquisitely complex task of combining bodily motion, balance, and controlled falling into standing and walking at will. The physics and calculus required to model this task for a robot are highly challenging to this day, even after decades of engineering research dedicated to that purpose.

Although the human ability to create, refine, and apply conceptual models is formidable, and has helped us to survive and prosper through the ages, it has two significant limitations: our models are very difficult to accurately communicate, and we are not good at predicting the behavior of a system when it includes a feedback loop. Both of these limitations can be powerfully addressed using mathematics, formal logic, and computer software.

The limits of fully communicating our conceptual models often become apparent when we initiate the *because-I-said-so* "sub-model," or someone uses it with us. Closest to home for a parent are the continual cases in which attempts to communicate our models of, say, good nutrition or personal hygiene are reflexively challenged with the question, "Why?" In such cases, *because I said so* may suffice as a functional sub-model for practical purposes. This same "sub-model" also is invoked by many professional practitioners, albeit using more customer-friendly phrasing, when a conceptual model is too nuanced to communicate. The physician, the attorney, and the financial advisor all rely on models of their technical domains, developed through formal education and experience, to effectively advise their clients. They pronounce a diagnosis, or declare a legal strategy, or propose an investment plan based on a tacit agreement that, for the most part, these are valid recommendations *because I said so*.

Most conceptual models, whether professional or personal, are not developed using mathematics and are not shared with others as precise, formal descriptions. These models are created primarily through training of the neural network hosted within our brains to discover and retain associations of information (e.g., a patient's symptom) with an explanation or response pertaining to it (e.g., a diagnosis and treatment). However, while we can help someone to understand a model through observation, trial and error, and repetition, it is very difficult to explicitly and fully communicate a model, even for something as seemingly simple as distinguishing a cat from a dog. It is simply more practical to repeat the training course for each of our children, complete with illustrations and repetitions, than it is to prepare a definitive documented model of comparative animal morphology and behavior.

Unfortunately, that practical approach is not sufficiently rigorous for the effective documentation and transfer of knowledge according to the scientific method. It is likely that few center fielders could formally document the process of trying to snag a 385-ft blast heading toward the fence, and most people probably cannot clearly explain their conceptual model for distinguishing a dog from a cat. The physician or attorney may have more success describing the model for a recommended treatment or a legal strategy since their professions depend to a great extent on systems of logic, but few could provide a succinct model for applying their professional judgment given all the changing assumptions and variables each case imposes. Inevitably, people with expert knowledge are faced with a gap between what they know "in their bones" and their ability to convince their clients or colleagues of its validity.

Fortunately, we have formal logic, and derivative mathematical and symbolic forms of it, which can help guide us from what is known toward new information as yet unknown. Formal logic provides uniform standards for reasoning and critical thinking, and accepted methods for applying logic are invaluable for documenting the validity of thought processes and identifying fallacies in them. Furthermore, powerful quantitative tools such as statistics, matrix algebra, and calculus extend formal logic into highly abstract realms of mathematics and science that are otherwise impossible to penetrate. These tools and their underlying systems of logic have made it possible to capture professional expertise as highly explanatory models of physical systems. Using various computer programming languages, expert knowledge can be encoded to create automated tools that simulate the consequences of altering the system's assumptions, parameters, and variables. These *simulation models* are also very good at processing the effects of feedback loops in systems—something of which the human mind is much less capable because the results may appear too complex for comprehension. An additional benefit of simulation models is that they produce repeatable results.

As indicated at the beginning of this text, however, modeling has generally remained a "sandbox" for computer programmers and other researchers who have the skills to translate conceptual models into mathematical algorithms, and then into computer programming languages. Practitioners who possess both subject matter expertise and excellent model-building skills have been almost as rare as alchemists; the communication gulf between scientist and modeling team has been almost as wide as the one between, say, attorney and client. The necessary division of labor between the scientist and the modeling team's mathematicians, statisticians, and programmers has encumbered both model development and adoption by interested practitioners. Why is this so?

First, creating models has typically been time-consuming and costly because of inefficiencies inherent in translating one expert's intuitive knowledge of system dynamics into the precise language of algorithms, and then translating the algorithms into computer code. Each translation has historically been executed by a different expert. In order to ensure that nothing has been "lost in translation," several iterations of revision and verification may be necessary before all contributors are confident of the model's scientific content and operation. The process competes with many other research activities for sufficient funding, personnel, and time allotted to produce results.

Second, even after the model has been verified and formalized by the development team, there is still the matter of validation (i.e., proof of accuracy). Without validation, field ecology peers who cannot readily look inside to evaluate it for themselves often regard the model as a "black box." On one hand, the prospective user deserves to know whether the model is accurate; on the other hand, the subject matter expert is put in a position of being considered guilty until proven innocent by second or third opinions—not unheard of in the professional world, but quite far from standard practice, too.

An unfortunate implication of this dilemma is that the highly technical aspects of model development have contributed to the alienation, or perhaps intimidation, of practitioners in the field who might greatly benefit from using explanatory simulation models. However, technology is on the side of scientists who have an interest in simulation modeling but no practical way to use it. The growing availability of open source software tools, and methods for using them to capture expert knowledge of system dynamics, now make it possible to develop spatially explicit simulation models without formal training or programming skills.

This book is written for ecologists and students of ecology who are interested in the idea of capturing and sharing their own undocumented conceptual models of natural systems using simplified software tools and proven collaborative methods of development. The immediate benefit of creating this type of model is that internal expert knowledge becomes clarified and quantified as a decision-support tool the scientific content of which may be reviewed by others with related expertise. These models may be revised and extended with relative ease, and some may even be adapted or repurposed beyond their original intent without starting again from a blank slate. The first part of the book summarizes current state of the science and art of ecological simulation modeling. It includes chapters dedicated to a survey of landscape modeling environments for users with no formal programming experience and methods for managing a multidisciplinary ecological modeling project. The second part of the book documents 11 case studies where expedient ecological simulation models have been developed and applied by university graduate students. These applications are used to support management activities ranging from species at risk and nutrient flows in rivers to food distribution and social services.

The editors of this collection have two objectives for the book. The first is to outline an expedient and effective methodology for fielding useful ecological simulation models. The second is to inspire readers with the confidence that it is within their grasp to create and use computer-driven tools that help to clarify and extract their professional expertise, and share it with their community of peers. We sincerely hope that this guide contributes to the advance of effective environmental management practice for the purpose of improving ecological sustainability.

Reference

Wilensky U (1999) NetLogo. Computer software. Center for Connected Learning and Computer-Based Modeling, Northwestern University, Evanston, Jan 2011. http://ccl.northwestern.edu/netlogo/. Accessed 01/2011

Chapter 2
A Collaborative Process for Multidisciplinary Group Modeling Projects

James D. Westervelt and Bruce Hannon

2.1 Introduction

Although a virtually unlimited number of ecosystem management issues may be illuminated using small, expedient simulation models developed by one person or a few, there are many cases where much more complex problems can be efficiently modeled by a relatively larger team that spans several disciplines or technical domains. This chapter describes a process for conceiving, coordinating, and launching such a model development initiative using the same simple software platforms described elsewhere in this book.

We have successfully taught ecosystem modeling to groups of university students as a three-stage process, with a sequence of steps comprising each stage. By adhering to this process, our students have developed highly utilitarian ecological simulation models that are based on real-world data and specialized technical expertise. Many of the models documented in this book were developed by students. Our model-development process can be outlined as follows:

1. Prepare to model

    (a) Identify objectives and scope
    (b) Identify available resources (personnel, expertise, time, software, hardware)
    (c) Consider group dynamics (including ownership issues)

J.D. Westervelt (✉)
Construction Engineering Research Laboratory, US Army Engineer Research
and Development Center, Champaign, IL, USA
e-mail: james.d.westervelt@usace.army.mil

B. Hannon
Department of Geography, University of Illinois,
220 Davenport Hall, 607 S Mathews, M/C 150, Urbana, IL 61801, USA
e-mail: bhannon@illinois.edu

J.D. Westervelt and G.L. Cohen (eds.), *Ecologist-Developed Spatially Explicit Dynamic Landscape Models*, Modeling Dynamic Systems,
DOI 10.1007/978-1-4614-1257-1_2, © Springer Science+Business Media, LLC 2012

2. Construct the model

   (a) Set modeling constraints
   (b) Conceptualize the full model
   (c) Design submodels
   (d) Construct a full dummy model
   (e) Construct the submodels

3. Integrate the model

   (a) Debug NetLogo[1] compilation errors
   (b) Debug errors in model logic
   (c) Demonstrate to end users

This process applies irrespective of the modeling software environment that is used. Each of these steps is discussed below.

2.2 Prepare to Model

2.2.1 Identify Objectives and Scope

The first step for successful multidisciplinary dynamic modeling efforts is to review the objectives and scope of the project. This involves answering the following essential questions:

- *Who is the end user?* This question is often glossed over because model developers often assume that the end user is exactly like them. A model's utility can be significantly limited when the needs and applications of a prospective external community of users are not fully considered at the outset. This often becomes clear to the model developer only when the model is demonstrated to others.
- *What does the end user require of the final model?* Determining what the identified end user expects from the modeling effort requires substantive communication among the modeling team and the end user. The answers to this question are not always immediately apparent to the prospective end user, and they often need to be drawn out through direct and open communication between modelers and prospective end users.
- *How accurate do the output requirements need to be?* There is no single correct answer to this question. For some end users, it may be sufficient for the model to be capable of generating relative or suggestive output to show trend directions. Other end users may require highly accurate data for purposes of fine-tuning land

---

[1]This text assumes that the modeling group is using NetLogo (Wilensky 1999) as its collaborative development platform.

management scenarios. Again, the most suitable answer is arrived at through direct, substantive communication between the modeling team and prospective end users.

2.2.2   Identify Available Resources

This task also can be aided by answering certain essential questions related to personnel capabilities and availability, data requirements, and computer technology, as discussed below.

2.2.2.1   Personnel Capabilities

- *What expertise is available through the modeling group?* If the group has the right mix of collective experience, model development can proceed rapidly and with great efficiency.
- *What actions are needed to fill any knowledge gaps in the team?* Depending on the gaps identified, the group may either need to recruit outside expertise or identify supplemental training needs for various team members.

2.2.2.2   Participant Availability

- *How much time will each participant be able to provide the modeling effort?* Regardless of how fast and proficient each team member is, and how well organized the team is, a participant's lack of availability during critical "windows of opportunity" for collaboration can delay progress. Availability of team members is easy to overlook because it may seem imponderable at the planning stage.
- *What time frames are available for inter-team coordination?* For interdisciplinary efforts, the viability of a group member must be assessed with respect to how much time the member can be available to coordinate with the others.

2.2.2.3   Available Data

- *Have other models already been developed for the subject environment?* Find out what other models are already available and how they may apply to the current problem. Investigate what existing models can teach the team about modeling the environment and about problems or limitations yet to be addressed.
- *What data are available?* Determine whether the model can simulate the subject environment to the required level of accuracy using readily available data.

If insufficient data are available to produce the level of accuracy needed, the responsible person needs to determine whether it is feasible to obtain more funding or work time to fill the data gap through additional fieldwork or research.

• *What is the full set of data requirements for the model?* Thoroughly consider all forms of data needed, not just technical information about the biome. A spatially explicit model requires spatial data such as raster maps, satellite imagery, vector maps, polygon maps, and point data (representing the location of data sampling points).

2.2.2.4 Computer Technology

• *What computer hardware is available?* Review the inventory of hardware available to support the project locally. Consider whether local network or Internet access is needed, and whether the available connections have sufficient bandwidth to support communication between connected machines.
• *What software capabilities are available?* In addition to selecting the software modeling environment, consider what ancillary software may be needed to support the modeling effort.
• *Do the benefits of the identified computer technology justify its cost?* It may be tempting to acquire sophisticated information technology that requires heroic effort to apply. This temptation may arise from the desire to do the best possible job, or simply to extend the joy of modeling as long as possible. From a business standpoint, this approach may not provide a sufficient benefit, so a less expensive approach (in terms of money, time, or both) may be optimal.

2.2.3 Consider Group Dynamics

Interdisciplinary efforts have special management needs. Traditional management hierarchies must be deemphasized in favor of more unmediated coordination and cooperation among team members. Individuals from different disciplinary backgrounds often base problem-solving on different paradigms, which makes some difficulty in communication likely. Such difficulties, however, provide an opportunity for significant "cross-pollination" of knowledge within the team. Focusing specifically on group management considerations that pertain to model building, several issues that have a direct bearing on integrating individual efforts into the group product must be recognized and addressed.

2.2.3.1 Model Development and Integration Responsibilities

Large, complex, multidisciplinary modeling efforts undertaken by a team must be split logically into model subcomponent development tasks. The role of each task in the project is defined in terms of the scope of the entire project. Each model subcomponent is completed through individual effort, imagination, and expertise. Development of the subcomponents begins with the definition of subcomponent requirements, continues with coding that is often accomplished in isolation, and completed with refinement efforts during full model integration. It is crucial that the owner of each subcomponent be involved during all phases of integration. An individual's temptation to "turn over" a model component should be discouraged until the entire project is completed. If a component that has been "completed" prematurely and ultimately needs refinement or revision based on overall model development, it is almost always more efficient for the author of that component to make the changes. In cases where it is necessary to assign a different team member to make the revisions, it is often better to completely rewrite the subcomponent than to spend much time trying to understand the original. For the sake of time, cost, pride in accomplishment, and avoidance of disruption, individual responsibility for subcomponents and submodels must be maintained throughout entire project.

2.2.3.2   Scope of Subcomponent Development Efforts

Each individual developer must understand that his or her task is the development of a submodel that provides essential outputs for use by the full model, not a fully functional model in its own right. Developers also should understand that they need to meet, but not exceed, the functional requirements of the subcomponent. A subcomponent that does significantly more work than is needed by the full model can drain developer resources and also produce a model that runs slower than what is acceptable.

2.2.3.3   Scheduling

People work best within the context of achievable expectations. Schedules define expectations from the perspective of available time, energy, and ability. A team must work off a common schedule that sets achievable goals within realistic timeframes. A feasible yet challenging schedule must be developed with and clearly communicated with the team. "Plan the model and model according to the plan."

2.2.3.4   Leadership Among Equals

A nonhierarchal, multidisciplinary team can be difficult to coordinate. As a practical matter, someone on the team should be responsible for pulling the team together as a working unit based on his or her understanding of everyone's individual personalities, expertise, and motivations. When a difference of opinion cannot be resolved by group consensus, someone must provide leadership in order to move forward. The designation of an effective team leader may be the most important decision made during the entire modeling project. Ideally, an effective team leader will be able to work well with all members of the team on both a personal and technical level. The person must fully understand the modeling team's overall objective, be well versed in the modeling process, and be trusted by the group to exercise impartial, well-informed judgment to avoid or resolve conflicts.

2.3 Construct the Model

2.3.1 Set Modeling Constraints

2.3.1.1 Potential Model Components

A large array of model components is available to the ecological modeler. Decisions made in the choice of hardware and software, however, will limit the range of options available. Typical model components to consider are briefly described below.

- *Landscape patches.* Landscapes are now recognized to be important variables in models that simulate the movement of individual organisms, structural changes in ecosystem boundaries, or the movement of air and water. Dividing the landscape into grid cells, hexagons, or irregular polygons is an effective way to capture some spatial information. For the sake of simplicity, most models are designed to use, at a fixed resolution, data in only one of those landscape patch formats. Landscape resolution is often chosen with respect to the operational time step incorporated for simulations.
- *Linear objects.* Some spatial structures are most efficiently simulated as linear objects. Examples include streams, rivers, and most of the built environment such as roads, fences, buildings, and parking lots. Incorporating both linear objects and landscape data stored as rasters or hexagons is complex, but it is often unavoidable. Just as different data storage techniques are appropriate to different data formats, so are different modeling techniques appropriate depending on the model output requirements.
- *Discrete mobile objects.* If individual people, groups of individuals, vehicles, or individuals of an endangered species are key variables in the model, they must be represented as discrete mobile entities. Such objects must be able to disconnect from a location in the landscape space and "reconnect" in an adjacent space to interact with the environment there.

2.3.1.2 Potential Model Interactions

There are a very large number of potential interactions among model components, so the following list indicates only broad categories of interaction. The modeling team must identify the model interactions that will best simulate the system.

- *Raster geographic information system (GIS) interactions.* Classes of traditional GIS interactions have been developed and discussed by Tomlin (1990). These include the following:

  - Simple location-by-location overlays that can be expressed with mathematical equations, using maps as variables. These can be used to find locations that meet certain local criteria; find correlations or potential local impacts related to a proposed change to the landscape; or transform a set of maps (e.g., slope, aspect, soil characteristics, rainfall) into new interpretations using mathematical relationships (e.g., soil erosion potential based on the Universal Soil Loss Equation).
  - Near-neighborhood operations at computer output as a function of the state of small regions surrounding each map location. These can be used to compute slope or aspect as a function of the elevations surrounding each map location, determine direction of a flow such as water or air, or identify areas where information changes rapidly (e.g., edge detection).

- *Cellular automata interactions.* Individual cellular automata (CA) comprise a two-dimensional surface that changes over time. The changes are driven by equations that generate the future state of each cell (i.e., location) as a function of the current state of that cell and its nearest neighbors (either four or eight depending on system design). For exploring the qualities and characteristics of the cellular automata approach, cells are often assigned to a limited number (fewer than 256) of states. Relaxing this limit to accommodate a large number of variables results in the type of models described in this book: models with fixed time steps that run simultaneously for a number of land parcels arranged in regular grid cell arrays. Each cell is treated as a homogenous system that can be influenced in each time step by its own state and the state of all its adjacent neighbors.

- *Vector GIS interactions.* Landscape information stored as polygons and linear features can also represent objects that interact and move such as traffic flow patterns along roadways or the hydrologic activity of stream or river networks during unusual storm events. Entities like cities, parking areas, private land, or stable ecosystem regions are most efficiently stored as polygon data elements, and they can be conceptually easy to model as distinct entities.

- *Mobile object interactions.* Some models require distinct entities that move across the landscape. Examples include individual members of an endangered species, vehicles moving about a landscape, or a group of individuals that moves collectively in close geographical contact.

These broad classes of interaction can combine into compound or hybrid types of interaction. Animals modeled as mobile objects must interact with water found in streams that are modeled as vector entities, and also with vegetation that is modeled as a component of a cellular model fixed in space. Entities can communicate with each other through sounds, propagules, pheromones, and waste gases that disseminate through the modeled space. It can be seen that the range of potential interactions is quite large. Generally, the group's options for modeling will be greatly constrained by the selected modeling environment and available computer hardware.

2.3.1.3 Simulation Timeframe

Once the purpose of the model has been determined, it should be relatively straight-forward to identify the optimal timeframe for simulations. The timeframe is determined on the basis of informed guesses about how rapidly the predictive power of the model will decay over simulated time. This same issue impacts the accuracy of the sophisticated large-scale models used to forecast weather patterns; even state-of-the-art models lose predictive power the longer the simulation runs. There may be exceptions to this, however. For example, spatial models may show stability at a gross scale while showing apparently random output at a detailed scale. That is, the overall pattern may remain the same over time, but details about the location of the pattern may change with different simulation runs. In general, the timeframe is determined directly by end user requirements.

2.3.1.4 Time Step Options

The simulation proceeds from a given starting point to the end of the timeframe identified above. Simulation time may proceed according to fixed time steps, variable time steps, or the occurrence of specified events.

- *Fixed.* This is conceptually the simplest approach, but functionally it is the most limiting. The model runs with a set time step, such as 0.25 or 1, which can represent days, months, years, or other measures of time. A known time step simplifies the model because all equations are generated with respect to the same time step. A fixed time step, by definition, cannot accommodate variability in the system. If the time step is weekly, for example, then the model cannot capture daily changes in temperature, moisture, or plant growth. Similarly, a daily time step would miss the effects of a flash flood that may take only minutes to cause great devastation to vegetation.
- *Variable.* There are two options available for using a variable time step. One option is to set the time step to be long, initially, but also allow it to be modified dynamically as changes occur and are detected within the model. In the case of the flash flood example given above, when the storm occurs the model would detect rapidly changing activities, stop the simulation, back it up, and resume the simulation using an appropriately smaller time step. The second option is to assign different fixed time steps to different parts of the model. This approach requires less computational power while maintaining the relative simplicity of fixed time steps.
- *Event driven.* This approach advances time not by steps but according to a calendar that schedules specific events. A plant submodel may execute plant growth and then schedule itself to be updated at some later time based on its own rate of activity. A storm submodel would be programmed to run at a specific time, and while it runs it can interact with the plant model and schedule the plant submodel to accelerate growth in response to the influx of water. This approach is most

attractive for models designed for limited computing resources. From a modeler's perspective, however, it is the most time-consuming approach because it requires significantly larger simulation models than are needed using the other two approaches.

As can be seen, one of the main criteria for selecting an approach is the capabilities of available hardware and software. When this criterion is not critical, the selection will be made on the basis of the costs and benefits of the alternatives.

2.3.1.5 Spatial Resolution Options

The resolution of the landscape's spatial representation must be considered in terms of the modeling objective, model subcomponents, and technology constraints. Two major aspects of dynamic, spatially explicit ecological models are the spatial distribution of model components and the effects of the space on model component interaction. The resolution of space in this type of model is as important as the resolution of time steps. Most modeling environments divide space into a checkerboard-type grid surface using cells of uniform size. If the cell size is too large, the implications of the spatial arrangement of objects in the system can be lost. As cell size is reduced, more computational power is required to support the model. Another consideration is how the cell size relates to agent activities and time steps; without a logical correlation among the three, the task of developing the model logic becomes much more difficult. In the case of an animal that covers up to 1 ha in a time step, for example, a cell size smaller than 1 ha will require the modeler to track the several cells comprising the specified 1-ha area and to ensure that the animal agent interacts with all of those cells during the time step. The development of the modeling logic to address that situation could be avoided simply by assigning the cells a size of 1 ha. That said, there is no universal formula for assigning cell size because other factors may equally influence the landscape surface scaling decisions. Three general terrain resolution schemes may be applied to simulation model design:

- *Fixed.* The terrain features assume a fixed resolution that can be constant across the landscape representation. A regular array of square grid cells or hexagons is commonly used in spatial simulation environments. These have the advantage of being conceptually simple as models need not account for different or changing resolutions.
- *Hierarchical.* Models that simulate activities occurring at different spatial resolutions may adopt a spatial data structure that maintains information in a hierarchical manner. Each cell or hexagon can be iteratively decomposed into increasingly smaller components. Large entities (e.g., weather systems, flocks of birds, clouds of spores or pollens) can move rapidly across the system using relatively long time steps and large spatial patches. Smaller entities (individuals or vehicles) can operate at smaller time steps and smaller patches. In this scenario data are maintained simultaneously at varying scales.

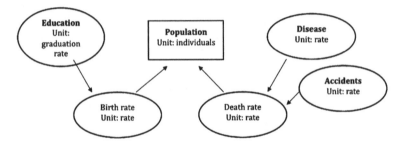

Fig. 2.1 Early state of model concept

- *Variable.* In the case of large objects that move slowly across the landscape (eco-systems, roaming herds of ungulates, or populations of invading species), there may be advantages to maintaining the entity as a single whole while retaining detailed spatial structure that defines the dynamically varying extent of the entity. This type of operation requires maintenance of the spatial extent at a fine resolution while allowing simulation of the object dynamics at a grosser resolution.

As is true with other considerations discussed above, the modeling staff will generally find their options constrained by the software and hardware limitations.

2.3.2 Conceptualize the Full Model

The full model concept begins with a focus on the model output requirements. These will almost always take the form of time series output showing the dynamic status of something in the model. The output series might track the status of an endangered species, property values, soil depth, land use patterns, ecosystem health, or the position and paths of vehicles or individual organisms. In all cases, a firm set of output requirements must be specified, and they will drive all subsequent decisions. First the outputs are documented as diagrams on paper, whiteboard, or computer screen (see Fig. 2.1). These represent the model's first *state variables*, parameters whose values will change over time. Each variable must be associated with a unit of measure. Some units will be straightforward (e.g., dollars, weight, mass, count), but others may be challenging. For each of these output variables, the group identifies the factors that directly influence the state of the variable. The diagram is revised as these factors are captured, with arrows showing the state variables that are affected.

At this point the group may have diagrams of system components with arrows as in Fig. 2.1. Boxed components are state variables; circled variables with arrows coming to them are calculated at each time step; and circled items with no arrows coming to them are model parameters (i.e., fixed variables). Equations will be written for every component in the diagram that has incoming arrows, so it is

important to begin ensuring that the units associated with the connected components are compatible.

This conceptual model grows over time. Participants may become anxious about
whether there is enough time, funding, energy, or information to complete the
model. Once a model is captured as a computer application, sensitivity experiments
can identify which model values, components, equations, and logic affect model
outputs to greater and lesser extents. It has often been repeated that a model is done
when nothing else can be taken out of it. Although the model remains conceptual at
this point, it is important that developers continually perform conceptual sensitivity
analyses to discover which model components are clearly not likely to significantly
affect model outputs. With judicious pruning and reference to the core project objectives, the conceptual model can be kept within limits that allow its completion using
the available time, energy, and expertise. The process proceeds by looking at the key
input variables that might be used to compute the state of the primary outputs
desired; team members look in turn at each of these variables and determine key
inputs required to compute their changing states. The group accomplishes this task
iteratively until the participants are comfortable that the conceptual model is sufficiently complex to answer the primary questions of importance to the end user, but
is still achievable within the identified constraints.

To summarize, the following steps are required to fully conceptualize the
model:

1. Identify and discuss the primary output desired. Begin with the questions that
   will be asked of the completed system.
2. Discuss the model inputs that might be required for generating accurate state
   changes in the primary state variables.
3. Perform a conceptual sensitivity analysis for each potential input to help prioritize them.
4. Repeat the previous two steps iteratively for each important input until the model
   is agreed to address all primary end user questions and to be feasible to complete
   given the availability of project resources.

2.3.3 Design Submodels

If the model is sufficiently large, development efficiency will require that it be
divided into submodels. Consider the design of a new automobile engine. The process may involve the expertise of dozens of people and it would be inefficient for all
of those people to collaborate in one room on the design of every component.
Instead, the design group is divided into smaller units with specific expertise to
independently design engine subsystems, such as fuel delivery, ignition, cooling,
exhaust elimination, and noise management. Each subsystem team is given specific
parameters to ensure that the subsystem performs as specified, fits the available
physical space, and is properly supported by other subsystems. The steps explained
below will help guide the effective development of submodels.

2.3.3.1   Identify the Submodels

Once conceptualization of the full model is complete, responsibility for the development of the submodels is assigned. Submodel teams then divide their responsibilities among individuals, who clearly understand their tasks in the context of group needs and the overall requirements of the final model. Each team retains primary responsibility for its submodels throughout the life of the project. Thus, this decomposition of the full model must be accomplished with close attention to team member:

- Expertise
- Availability
- Learning requirements

This stage often proceeds smoothly since the capabilities of each participant are often expressed during model conceptualization.

2.3.3.2   Set Submodel Requirements

This portion of the work starts with further conceptualization of the submodels, with the immediate objective being to identify submodel input requirement that will need to be satisfied by other submodels. This process becomes an iterative conversation among all submodel groups. Through identification of submodel input requirements and submodel output possibilities, the teams work out design and development contracts with each other. Stated concisely, the three steps are for teams to:

1. Identify external input requirements for their submodels.
2. Identify potential outputs of their submodels.
3. Agree on contracts with all other submodel teams based on the required inputs and outputs.

The contract must identify publicly available state variables, variable units, and variable resolution requirements. Debate among teams will likely proceed through several iterations before a consensus is reached. It is important that this debate and the resulting contract be taken as a firm obligation. If a team labors to develop a submodel based on the agreement that a key input will be available, then that input must be made available. Failure to meet these obligations can severely weaken team cohesiveness and jeopardize successful completion of the model.

Table 2.1 illustrates a representative, partially completed "submodel team agreement table." The column headings are filled with parameter name, associated unit, variable type, and—grouped at the right—the submodel team names. The row headings do not list all state variables in the model, but only those that are shared across models, which are those that are initialized and maintained by the submodels. The agreements captured in this table are (1) the variable names, (2) designation of the submodel responsible for each variable, and (3) the units to be applied. In some cases it is also important to include an agreement on the update frequency and the

Table 2.1 Submodel team agreement table

	Units	Type: S-state, V-variable	Weather	Reservoir	Agriculture	Population	Economy	Civil engineering
Temperature avg	Degrees C	V	O	I	I			
Temperature low	Degrees C	V	O		I			
Temperature-high	Degrees C	V	O		I			
Rainfall	Inches	V	O	I	I			
Cloud cover	Percent	V	O					
Insolation	Joules	V	O	I	I			
Res volume	Liters	S		O				
Res withdraw	Liters	V		I	O			
Food crops	Tons	S			O			
Population total	Count	S				O	I	I
Population child	Count	S				O	I	
Population adult	Count	S				O	I	
Population elder	Count	S				O		
Income average	$/house	V					O	I
Employment rate	Percent	V					O	O
Paved roads	Km	S						O
Dirt roads	Km	S						O

uncertainty of the value. The variable type column indicates whether the value is a state variable (decreased/increased each step) or a simple variable that is calculated at each time step. An "O" in the table is used to indicate that the value is an output of the submodel directly above it and an "I" indicates that the value is an input. The table is not completed until every variable not associated with an "I" is deleted and every other variable is associated with one—and only one—"O."

2.3.3.3 Model Identification Requirements

As a special case of the submodel requirements identification, model state variables must be initialized with a system state at time step 0 (zero). Depending on the approach to state variable simulation, external data sources such as raster and vector maps, site description tables, entity state descriptions, and external model output (e.g., a global climate model) will be required to seed the model's state variables. This effort can be as time-consuming and difficult as the development of the model rules and equations. Team members will be assigned to this effort and will similarly debate and establish working contracts with the submodel teams, as described in the previous paragraph.

2.3.4 Construct a Full Dummy Model

To facilitate the integration of submodels into a functioning whole, it is very useful to assemble a "dummy" model that contains simplified versions of all submodels. At a minimum at this stage, each submodel should feed static values of all variables that it is responsible for providing to other submodels. The dummy model serves as a sort of workbench for each submodel development team. Some useful dynamics may be built in as data are available. For example, a dummy weather model might report a fixed series of monthly average temperatures, rainfall, etc., for use elsewhere in the model; or a population model might generate a monotonically increasing total population count needed by another submodel. The goal is to provide a simple, clear model context within which submodel development can be accomplished. That context includes the "wiring" between all submodels, which allows each submodel development team to replace their dummy submodel with working versions while maintaining all expected submodel outputs and using all preestablished inputs.

As submodel development proceeds, developers are likely to discover that they need additional inputs not previously agreed to, or cannot generate certain agreed-upon outputs, or can easily provide potentially useful new outputs not previously discussed. Therefore, during development it is important for teams to communicate all revisions needed to the contracts captured in the "agreement table," and to generate updated dummy submodel components as appropriate.

2.3.5 *Construct the Submodels*

By this point the submodels have been identified; submodel development teams have been formed; submodel interaction requirements, expectations, and initialization requirements have been documented; and timeframes for completion of the submodel components have been set. Submodel development teams can now focus independently on the further design, refinement, debugging, and sensitivity analysis of the submodel components. If the identified submodels are found to be too unwieldy, they must be broken into manageable components using the procedure described above for decomposing the full model into submodels. A more ambitious model will require several levels of partitioning before it is suitable for individuals to focus on developing the cause-effect mathematics that drive the model.

There is no universal method for constructing submodels because specific objectives and individual modeler capabilities vary to a great degree from project to project. This stage of model development is where individuals can be most creative with respect to the modeling process. Nevertheless, a number of activities and objectives must be considered during a submodel design and development exercise. These may grouped into two categories: general modeling and group modeling. The general modeling category involves such considerations as keeping the model simple, making sure it can be understood by the intended audience, ensuring that units used in the model conform with the submodel team agreement table, and performing sensitivity analyses. Of more interest here are group modeling principles related to submodel design and development:

- The submodel must be developed within the parameters established for the project.
- Duplication of names for variables, stock, and other values that will be publicly visible (e.g., "age") must be avoided among teams in order to prevent subsequent difficulties during model integration.
- All modelers must use only the software and hardware designated for the project.
- Submodel development will depend only on inputs generated by other submodels.
- All outputs required from the model must actually be generated by simulations.
- All inputs and outputs are used and generated using the units previously agreed upon by the entire group.
- Submodel development is completed within the negotiated timeframes.
- All required changes are communicated quickly and diplomatically to other submodel teams.
- Submodel teams continue to monitor the internal state and external input variables for their submodel to determine whether the submodel is operating within reasonable parameters.
- Submodels are tested and initially refined using group-generated artificial time series data.

As the individual submodels are completed, they are ready to be integrated into the full model.

2.4   Integrate the Model

When multiple submodels are completed and ready to be joined, the process of integration is usually performed by a subset of the entire modeling team, but that subset should include a representative from each submodel team. Integration is accomplished by inserting finished submodels into the dummy model, one at a time, while testing and debugging them to identify and eliminate errors in program execution and output. Because "interesting" things are likely to happen when finished submodels start working with each other's actual output for the first time (instead of dummy outputs), the necessary problem-solving work may become vexing if it is not approached systematically.

The most effective general approach to testing and debugging is to add the finished submodels one at a time, first ensuring that NetLogo can execute them without problems, then testing one submodel's operation in tandem with another one. This approach is familiar to anyone who has tried to diagnose and rectify problems with a complex device or system—a desktop computer or an automobile, for example. The intent is to isolate and resolve one problem at a time in order to avoid complicating the diagnosis process with divided attention or irrelevant variables. Once the two submodels are executing correctly in NetLogo and producing output that is reasonable and realistic, then a third submodel may be inserted and tested.

Although submodel integration must proceed in a controlled, systematic way, it cannot be prescribed as a linear procedure. Integration is an iterative process. It may require repeated testing and debugging work with different combinations of submodels, or even taking a step back to fix a new problem with a previously operational submodel that appeared after debugging a different one. The successful isolation of problems involves detailed inspection of each submodel's behavior within the context of the evolving operational model, and debugging requires participation by representatives of all submodel teams. The difficulty of these tasks depends on the specific model and the complexity of the system behaviors it simulates. Some trial and error during the debugging and integration of submodels is unavoidable. As noted previously, debugging addresses the two basic issues described below: compilation and logic errors.

2.4.1   Debug NetLogo Compilation Errors

Each submodel was developed in relation to the common time series output test environment embodied by the dummy model, so the finished submodel should "plug into" the dummy model with few problems, if any. Nevertheless, when a submodel is first inserted in place of its corresponding dummy submodel, certain "mechanical" types of errors may emerge right away. We refer to these as *compilation errors*, and they are revealed in the form of NetLogo error messages. They represent obvious errors with respect to the dummy model, such as incorrect unit errors or duplicated variable names that mean different things in two or more submodels. Resolution of these problems is generally a straightforward task.

2.4.2   Debug Errors in Model Logic

As NetLogo compilation problems are resolved and operational testing continues, the submodels will be responding to input combinations not included in the dummy model (i.e., authentic input provided by other completed submodels). At this stage, it is common for test simulations to produce incorrect output or other incoherent behavior. Examples include unexpected cycling, chaotic activity, operation of submodels outside their range of sensitivity, and nonsensical output. These results indicate errors in the logic embodied in the submodels, particularly in terms of how each one interacts with the others. The testing and debugging of model logic draw on the technical expertise of the team and each working group's deep familiarity with its own submodel. Each team must provide guidance for evaluating whether inputs from other submodels are within the specified ranges. This can take the form of inserting code that produces an explicit error message when a submodel is not receiving valid inputs. These error messages can help to facilitate communication between the submodel teams, but successful debugging further requires that all submodel teams collaborate very closely to assess model integrity and to monitor how their own submodels perform within the context of the whole system. All requirement changes and fixes must be assiduously reflected in the shared model documentation.

2.4.3   Demonstrate to End Users

Once all issues with submodel performance and operation of the integrated model are resolved, the development team provides demonstrations for representative end users. The purpose of the demonstrations is to let the end users run the model and see how it performs when loaded with real or simulated data. Two reasons for providing demonstrations are to observe whether the target users have any unforeseen problems running the model and whether they offer any suggestions for last-minute "tweaks."

If the modeling team has followed our three-phase development process, no significant problems should emerge. Straightforward changes, such as level of detail displayed in the user interface, may emerge; these are easy to implement. Direct requests by users for more substantive revisions should be considered in terms of model utility and cost of implementation. Fairly substantive changes are not unheard-of after end user demonstrations, but at the final stage of a project the project team tends to be constrained in terms of revisions it can consider. The nonnegotiable constraint is the project objective: user change requests that exceed the original scope of the project can rarely be satisfied under the original funding and personnel commitments. However, constructive user input for significantly extending the scope or utility of the model should not be discarded: it may suggest the need for a follow-on project when funding and time are available, and may even provide core concepts for a more ambitious model development project in the future.

2.5 Disseminate the Model

Product distribution is not part of the model development process, strictly speaking. But dissemination of the model to a community of interest is usually the final, if unspoken, goal of nonproprietary model builders. Even given the level of interest and effort each member of the multidisciplinary team has invested to fulfill the specific modeling objective, the final product may have important potential uses not recognized by the team. Transparent, open-source models such as those constructed using NetLogo may be readily adaptable for different applications within any of the disciplines represented by the modeling team. An operational model may serve as a sort of template—or at least as a concrete starting point—for specialized or enhanced follow-on models. Teams that wish to maximize the utility of their modeling effort will think about efficient modes of model distribution through the course of the project. Common modes of model dissemination include formal documentation and publication of the model using the *Overview, Design concepts, and Details* or ODD format (Grimm et al. 2006); posting the model and user documentation to appropriate user-community web sites; and distribution on physical media such as CD-ROMs.

2.6 Conclusion

This chapter has outlined a practical framework within which an interdisciplinary research team can design and develop a large, dynamic, spatially explicit ecological landscape simulation model. This framework, repeatedly used with success with our university students over many years, is intended to promote an effective balance between efforts that need approval by the whole development team and efforts that draw on the imagination, expertise, and motivation that arises through the effort of the individual. Full model conceptualization is performed with respect to end user requirements, as tempered by available resources. That conceptualization governs the partitioning of the model into components that can be developed through individual efforts. As the model components are completed within the overall design requirements, they are linked together as an operational final model. Full model debugging requires modification to the components by the original developers of those components in coordination with the developers of the other components.

Quality leadership in this multidisciplinary environment is crucial for success, and it must be executed with consideration and respect for differences in the personality, background, motivation, and time availability of the team members. Using the approach outlined in this document will help the team leader and members to successfully design, develop, and operate large, complex spatial models.

References

Grimm V, Berger U, Bastiansen F, Eliassen S, Ginot V, Giske J, Goss-Custard J, Grand T, Heinz SK, Huse G, Huth A, Jepsen JU, Jørgensen C, Mooij WM, Müller B, Pe'er G, Piou C, Railsback SF, Robbins AM, Robbins MM, Rossmanith E, Rüger N, Strand E, Souissi S, Stillman RA, Vabø R, Visser U, DeAngelis DL (2006) A standard protocol for describing individual-based and agent-based models. Ecol Model 198(1–2):115–126
Tomlin D (1990) Geographic information systems and cartographic modeling. Prentice-Hall, Englewood Cliffs
Wilensky U (1999) NetLogo. Computer software. Center for Connected Learning and Computer-Based Modeling, Northwestern University, Evanston. http://ccl.northwestern.edu/netlogo/. Accessed 01/2011

Chapter 3
An Introduction to the NetLogo Modeling Environment

David Stigberg

3.1 Background

3.1.1 Program Overview

NetLogo (Wilensky 1999) is a user-friendly agent-based modeling environment created by Dr. Uri Wilensky at the Center for Connected Learning and Computer-Based Modeling (CCL), Tufts University, Medford, MA. In 2000, the CCL moved to Northwestern University, Evanston, IL, where NetLogo development has continued to date. This discussion is based primarily on NetLogo 4.1.

NetLogo runs on most desktop computer platforms, including Microsoft Windows, Apple operating system OS X, and Linux. The modeling environment is programmed mostly in the cross-platform Java language (Oracle Corp., Redwood Shores, CA), but the BehaviorSpace component of NetLogo and the user-code compiler are written in a language called Scala. The language actually used to develop simulation models is called NetLogo, which is a "dialect" of the language called Logo, as is the well-known derivative language called StarLogo. NetLogo differs from both of those in many respects, particularly in terms of ease of use and greater power. Interested readers can find details about this in the NetLogo frequently asked questions (FAQ) document and the Programming Guide, both of which are available on the NetLogo web site (http://ccl.northwestern.edu/netlogo/).

This chapter is written to introduce new users to the NetLogo modeling environment, particularly people who have no substantive exposure to simulation modeling but understand the benefits of capturing their personal expertise as a dynamic,

D. Stigberg (✉)
U.S. Army Engineer Research and Development Center, Construction Engineering
Research Laboratory, 2902 Newmark Drive, Champaign, IL 61822, USA
e-mail: David.K.Stigberg@usace.army.mil

J.D. Westervelt and G.L. Cohen (eds.), *Ecologist-Developed Spatially Explicit Dynamic Landscape Models*, Modeling Dynamic Systems,
DOI 10.1007/978-1-4614-1257-1_3, © Springer Science+Business Media, LLC 2012

spatially explicit model. Despite its high degree of accessibility to novice modelers, however, NetLogo also may be very useful to others who have more demanding requirements, such as researchers experienced with higher-power modeling technology or those who wish to extend NetLogo output by linking it to a geographic information system (GIS), statistical analysis software, or a specialized mathematics package. This text focuses on fundamentals of using NetLogo, which was used to develop the models presented in Part II of this book, but it also points to supplementary information intended for users with more demanding modeling ambitions.

This chapter addresses the following aspects of NetLogo's suitability for the prospective user and his or her intended use:

- Ease of software use, particularly in terms of working through the interface and writing code.
- Scope of model-building features and effectiveness of NetLogo's implementation of them.
- Quality of software documentation, tutorials, sample models, and other support.

The chapter opens with an overview of NetLogo's origins and its technical capabilities.

3.1.2   Capabilities and Features

NetLogo can model both mobile and immobile agents. It handles multiple agents occupying the same physical space, and can include thousands of agents in a single simulation. Other capabilities include agent linking and networking capabilities, both two- and three-dimensional (2D and 3D) display, and easy switching between single-step execution and continuous simulations.

Features that distinguish NetLogo include:

- Clear and thorough documentation, including a large and rich set of sample models to instruct and stimulate the would-be modeler.
- An easy-to-understand graphical user interface (GUI) through which a variety of easy-to-use tools are accessible for shaping a model, controlling it, and monitoring its behavior.
- The easy-to-learn NetLogo modeling language, which is a powerful and flexible tool for model creation.
- HubNet, a "participatory simulation" tool that enables multiple users running separate client programs to interact with a NetLogo model.
- Tools that facilitate running multiple simulations in the background.
- BehaviorSpace, a tool that facilitates running a model many times using different inputs to explore the consequences of alternate scenarios.
- Icon editors and import capabilities for agents and links.
- A separate environment with dedicated tools for graphically creating system dynamics models instead of agent-based models.

- The ability to easily convert models into *applets* (small Java applications) for web viewing and experimentation.
- An application programming interface (API) of controller functions that allows NetLogo to be embedded in other applications.
- Software extensions that augment NetLogo's basic capabilities, including a special API for custom development of new extensions.
- A substantial GIS capability, providing the ability to load both vector (point, line, polygon) and raster (gridded) geographic data layers, display them, and incorporate their real-world features into models.

3.1.3 Installation and Setup

NetLogo may be downloaded free of charge from the NetLogo website, http://ccl. northwestern.edu/netlogo/. The installation process is simple for computers running Windows, OS X, and Linux, and simply involves downloading an executable installation file, double-clicking it, and following the prompts. Once installed, NetLogo is ready to run any of the included sample models to help the new user become acquainted with the nature of agent-based simulations. You can examine and tinker with the sample models at your own pace, or else follow the well-documented tutorials to immediately start preparing to build your first model.

3.2 Description of the Modeling Environment

The top-level NetLogo GUI provides three principal work areas organized as tabs near the top of the screen: the *Interface* tab, the *Information* tab; and the *Procedures* tab. The Interface tab is where model development usually begins, and also where existing models are selected, run, and manipulated. The Information tab provides a space for reading and writing succinct model documentation comments. In NetLogo's many sample models, the Information tab contains user documentation that can be highly instructive for novice modelers. The Procedures tab is the workspace where the model's *source code*—the program—is developed and stored. This section discusses each tab in its order of placement in the NetLogo GUI.

3.2.1 The Interface Tab

Figure 3.1 is a screen shot of the NetLogo sample Wolf Sheep Predation model (Wilensky 1997) with the Interface tab selected, shown running on a personal computer. This model, which you can open and experiment with after installing NetLogo on your computer, consists of three elements: grass, sheep, and wolves. Immediately on launching this model, the model view occupying the right half of the Interface

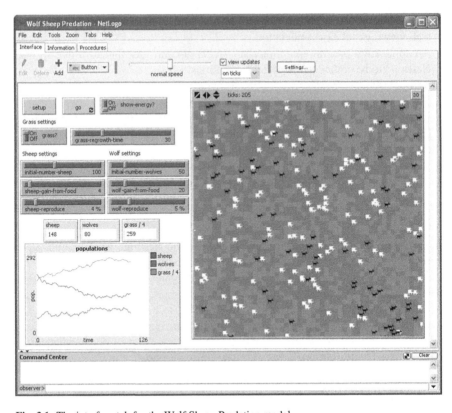

Fig. 3.1 The interface tab for the Wolf Sheep Predation model

tab is filled with a random array of live (green) and dead (brown) patches of grass. Within this field of grass, user-specified quantities of individual sheep and wolves are randomly placed. When the model is run at regular time steps, called *ticks* in NetLogo, the sheep and wolves are randomly moved. The consequences of the movement are then evaluated. Sheep may gain or lose energy depending on their movement to live or dead grass patches. Whether the sheep survive, reproduce, or die depends on their energy gain or loss, and also whether they encounter a wolf. Whether wolves survive, reproduce, or die also depends on their own energy levels, which depend on their ability to land on and eat sheep.

Immediately beneath the three tab labels, we see a horizontal toolbar with a core set of controls (buttons, sliders, and dropdown menus) that are activated or dormant depending on which tab is selected. With the Interface tab selected, the toolbar grays out any tools not applicable to the tab's functionality. The workspace includes a variety of additional tools immediately below, on the left half of the screen. The actual Wolf Sheep Predation "world" is displayed on the right half. At the bottom of the workspace is the *Command Center*, a field where commands may be keyed in to affect a simulation without affecting the model's source code.

In the Wolf Sheep Predation model, the world is displayed as a 51×51 square, 2D grid showing all model elements (i.e., grass, sheep, and wolves). In the horizontal bar at the top of the modeled space in Fig. 3.1, the simulation is shown to have already run for 205 ticks. The individual sheep and wolves are examples of mobile agents which, oddly, are called *turtles* in the jargon of NetLogo. The individual squares of grass, called *patches*, are examples of a stationary agent. The model world includes not only turtles and patches, but also an *observer* agent and, in many models, *link* agents that connect turtles in various ways.

Taking a closer look at the buttons and widgets displayed next to the model space in Fig. 3.1, we see the following, moving right and down from the upper left:

- A *setup* button, which sets the initial placement of sheep, wolves, and the living and dead grass patches.
- A *go* toggle button that starts and stops the simulation.
- A switch labeled *show-energy*, which specifies whether the current energy for each sheep and wolf is displayed.
- A switch labeled *grass*, which specifies whether grass status (i.e., live or dead) will affect the model's behavior.
- Six *slider* controls with self-explanatory labels, which the user manipulates to specify the initial values for various model parameters, including numbers of sheep and wolves, energy gains from food for each, and reproduction percentages.

Below these buttons and sliders are data display components that visualize information as the simulation runs. In the Wolf Sheep Predation model we see two types of these displays:

- Three *monitors*, which track the current amounts of sheep, wolves, and live grass.
- A *plot* with two labeled axes (population and time) showing changes in sheep/ wolf populations and grass cover[1] at each tick during the run time, which echoes the monitor values in graphical form.

The Command Center at the bottom of the workspace is for typing certain ad hoc commands directly to agents during a simulation, such as changing the display color of sheep or specifying which direction particular sheep should rotate. These commands are not added to the model's source code, and they do not persist after the simulation has ended.

At this point, it is important to interject that the NetLogo Interface tab will look somewhat different in each model. The general organization of the workspace is the same in all models, but the buttons and other devices that appear on the left side of the tab are custom controls added by the model developer. These will vary depending on the nature of the simulation model that is being developed. The basic model elements, for example, will vary in number and name, as will the types of input tools

---

[1] The value for grass is divided by 4 ("grass/4") so it scales to the plot display while still showing the trend in live cover.

provided for toggling certain conditions off and on, setting the values of parameters, providing dynamic data display during simulations, and so on.

Similarly, it is important to note here that the command buttons, toggles, and sliders do not become functional until they are referenced as variables within the NetLogo code. Certain types of ephemeral commands can be added using the Command Center, as explained above, but the actual model source code is developed in the Procedures tab, which will be described after a brief introduction to the Information tab.

3.2.2 The Information Tab

Figure 3.2 shows the Information tab for the sample Wolf Sheep Predation model. Commands for finding text in this tab (*Find*) and editing or writing it (*Edit*) appear as icons in the horizontal toolbar just beneath the three tabs near the top of the window. The content of the Information tab is unique to each NetLogo model, as it is in the Interface and Procedures tabs.

This example shows a concise summary of the model's nature and purpose, an explanation of what the model does, and a description of how to use the controls to run and control the simulation. Note that the text continues beyond the bottom of the window and is accessible using the standard scrollbar at the right side of the window. The model builder may use this tab to make notes about the model development process. When the model is verified and ready to distribute, the text in the Information tab may be further developed to serve as an instructional readme file targeted at new users of the model.

3.2.3 The Procedures Tab

The code-editing workspace for NetLogo—the Procedures tab—is shown in Fig. 3.3 for the Wolf Sheep Predation model. It is little more than a simple, specialized text editor designed for writing and revising computer code in the NetLogo programming language. The toolbar contains a Find command (functionally identical to the one in the Information tab); a *Check* command, a debugging capability that highlights errors in source code syntax; a dropdown navigation menu, which lists all the model's user-defined procedures; and a switch to toggle automatic indenting of the code. In the Procedures tab as it displays on a computer screen, the NetLogo programming language is color-coded. Commands are blue, *reporters* (commands that return a value) are purple, and other keywords are green, for example.

Most of the work to create a NetLogo model involves the writing and checking of computer code. But compared with many other computer languages, the NetLogo language is easy to comprehend, read, and write. It is an economical language because its vocabulary of commands is very well considered and focused on the

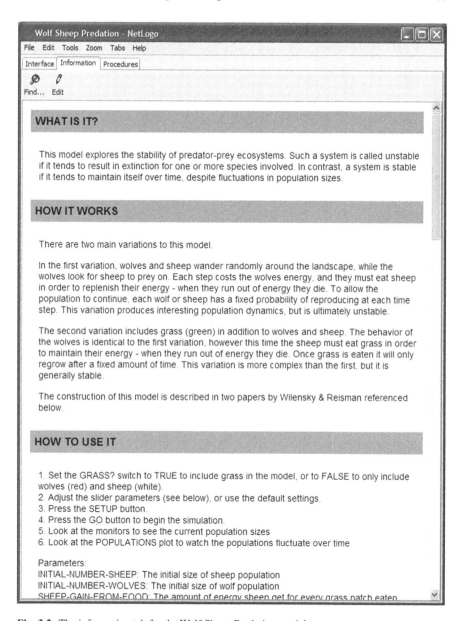

Fig. 3.2 The information tab for the Wolf Sheep Predation model

specific problems of model creation. Consequently, the imaginative application of a relatively small amount of NetLogo code can result in simulation models of considerable power and explanatory value. This fact is evident in the quality of the sample models that come bundled with the NetLogo application. In examining the Procedures tabs of these models, many new users will find that they can reuse portions of the source code for their own models.

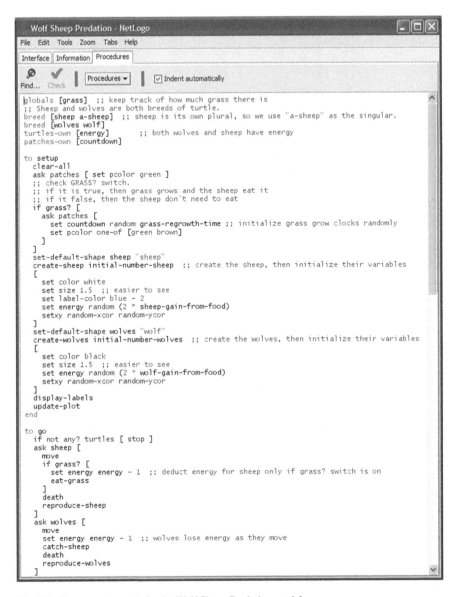

```
Wolf Sheep Predation - NetLogo                                      _ □ ☒
File  Edit  Tools  Zoom  Tabs  Help
Interface │ Information │ Procedures
🔍  ✓  │ │ Procedures ▾ │ │ ☑ Indent automatically
Find...  Check

globals [grass]  ;; keep track of how much grass there is
;; Sheep and wolves are both breeds of turtle.
breed [sheep a-sheep]  ;; sheep is its own plural, so we use "a-sheep" as the singular.
breed [wolves wolf]
turtles-own [energy]       ;; both wolves and sheep have energy
patches-own [countdown]

to setup
  clear-all
  ask patches [ set pcolor green ]
  ;; check GRASS? switch.
  ;; if it is true, then grass grows and the sheep eat it
  ;; if it false, then the sheep don't need to eat
  if grass? [
    ask patches [
      set countdown random grass-regrowth-time  ;; initialize grass grow clocks randomly
      set pcolor one-of [green brown]
    ]
  ]
  set-default-shape sheep "sheep"
  create-sheep initial-number-sheep  ;; create the sheep, then initialize their variables
  [
    set color white
    set size 1.5  ;; easier to see
    set label-color blue - 2
    set energy random (2 * sheep-gain-from-food)
    setxy random-xcor random-ycor
  ]
  set-default-shape wolves "wolf"
  create-wolves initial-number-wolves  ;; create the wolves, then initialize their variables
  [
    set color black
    set size 1.5  ;; easier to see
    set energy random (2 * wolf-gain-from-food)
    setxy random-xcor random-ycor
  ]
  display-labels
  update-plot
end

to go
  if not any? turtles [ stop ]
  ask sheep [
    move
    if grass? [
      set energy energy - 1  ;; deduct energy for sheep only if grass? switch is on
      eat-grass
    ]
    death
    reproduce-sheep
  ]
  ask wolves [
    move
    set energy energy - 1  ;; wolves lose energy as they move
    catch-sheep
    death
    reproduce-wolves
  ]
```

Fig. 3.3 The procedures tab for the Wolf Sheep Predation model

All source code for a model can fit in a single source file, and it is the content of that file that is displayed in the Procedures tab. However, there is an option to incorporate multiple source files into a model by using the _includes_ keyword. The purpose of this is to extend the basic model's functionality by exploiting NetLogo source code that may already have been written for a similar application. Other ways to extend NetLogo programming capabilities are discussed in Sect. 3.3.6.

3.3   Using NetLogo

3.3.1   Workflow Description

A model development project typically begins with the creation of Setup and Go control buttons using the Interface tab to establish a new NetLogo modeling workspace. Next, the modeler selects the Procedures tab to write the essential startup code for a rudimentary working model. This initial coding task includes the specification of the spatial grid within which the simulations will operate and the creation of mobile and stationary agents that comprise the working elements of the model. As the project proceeds, much time will be spent alternating between the Procedures tab to develop source code and the Interface tab to build the model user interface that links custom controls and output display monitors to the source code. The coding of new agents or revisions of spatial grid parameters, for example, will make it necessary to revisit the Interface tab to add or modify controls for parameter values, shapes, or other functionality.

Although model building is primarily an exercise in writing and debugging NetLogo code, the workflow can be surprisingly free-wheeling, almost gamelike. The coding of an agent or a new behavior via the Procedures tab may immediately propel the modeler back to the Interface tab to create a control or monitor tool for the user. NetLogo facilitates not only this sort of extemporaneous development procedure but also the executing of interim results so the modeler can get immediate feedback about his or her work. These characteristics greatly contribute to the attractiveness and accessibility of NetLogo to new modelers as they gain confidence and refine their models.

3.3.2   Coding in NetLogo

A comprehensive exposition of the NetLogo language is beyond the scope of this chapter. However, the examples provided below will provide the first-time user with an accurate sense of NetLogo's character and accessibility. As indicated previously, it is much easier to write agent-based simulation modeling code in NetLogo than in conventional general-purpose languages such as C++ or Java.

One reason for this ease is that the NetLogo language was expressly designed for the specialized purpose of building simple, dynamic, spatially explicit simulation models. The language contains no vocabulary, operators, or rules that do not pertain directly to building models, so there is less of it to master. It is powerful but economical, with an easy-to-grasp syntax, designed to shield the programmer from many of constructs that would have to be created manually in languages that operate at lower levels. This advantage may be better appreciated when examining the entire contents of Procedures tab for the Wolf Sheep Predation model, about half of which is shown in Fig. 3.3. The entire model contains only about 100 lines of source code (excluding programmer comments), many of which are only a handful of characters long.

Another reason NetLogo is easy to use is that developers have prebuilt many elements that any modeler will need on a recurring basis, eliminating the need to code them from scratch. Examples include the 2D and 3D grid spaces where simulations are staged. These are built into the programming environment, and their characteristics can be easily modified using the Interface tab without resorting to manual coding. The same is true of other stock graphical controls and monitors previously referred to. All of these would represent substantial manual coding tasks if using a general-purpose programming language, but they are ready-made in NetLogo so programmers can focus developing the application instead of the lower-level environment.

The accessibility of NetLogo coding skills can readily be seen in the following examples. The following three lines of code, extracted from Tutorial #3 in the NetLogo User Manual, comprise form a rudimentary setup procedure:

```
clear-all

create-turtles 100

ask turtles

    [setxy random-xcor random-ycor]
```

Each line of the code above is constructed using NetLogo *primitives*, which is NetLogo's term for programming elements that are defined by the language itself, as opposed to procedures that are created through the combination of primitives and other elements. Below is an explanation of each line:

- *Clear-all*: this is a general-purpose function that calls several other primitives to reset all global variables, reset the tick (i.e., time step) count, clear mobile agents, patches, plots, etc.
- *Create-turtles 100*: this line of code creates 100 new turtles, which are the default mobile agents in NetLogo.
- *Ask turtles…* : this is a command that "asks" all turtles to execute the list of directives that follows inside the square brackets; in this case, x and y refer to the axes of a 2D grid, and [*setxy random-xcor random-ycor*] directs the model to set the *xy* coordinate (location) of each individual turtle randomly.

The lines above apply commands to each turtle, irrespective of actual *breed* (i.e., type). To differentiate mobile agents, we designate separate breeds in the manner shown below:

```
breed [wolves wolf]
```
or
```
breed [sheep a-sheep]
```

To someone new to writing code, a detail of interest is that both a singular and plural term must be provided for each breed. For a case such as sheep, in which the singular and plural forms are identical in plain English, it is necessary to invent a singular term for purposes of writing unambiguous source code (a-sheep, in this case).

After differentiation of breeds, we then write code that creates the desired numbers of sheep and wolves:

```
create-wolves 50

create-sheep 100
```

Once the breeds are defined, we are able to substitute wolves and sheep (and an individual wolf or a-sheep) into any of the commands that govern turtles.

The following line uses a built-in command (*ask*) and points to a user-defined procedure (*move*):

```
ask wolves [move]
```

The user-defined procedures are identified by beginning with the word "to":

```
to move [

rt random 360

fd 1

]
```

which causes each wolf to rotate (rt) right randomly between 0 and 360°, and then move forward (fd) one unit on the grid. Once this procedure is created, the model can invoke it whenever it is needed.

3.3.3 Ease of Initialization, Compilation, Execution, and Error Messaging

As noted previously, NetLogo simulations are run from within the NetLogo environment using the Interface tab. First, the user *initializes* the model by clicking the Setup button. Initialization involves the launch of a user-programmed Setup procedure that returns the model to its intended default state prior to running a simulation. Next, the user clicks the Go button, which invokes a user-programmed procedure that executes all actions intended to occur in a single tick. Typically, this Go procedure consists of multiple user-created sub-procedures, and the Go button repeats ticks indefinitely until either the user clicks the button again or a condition written into the source code occurs. The compilation and execution processes that prepare the model to run happen behind the scenes and require no intervention by the user. Section 3.3.4, below, provides some basic information on running NetLogo models independently of the NetLogo environment.

Syntax and other coding errors that cause runtime errors during program execution will interrupt the simulation. NetLogo will highlight the incorrect code and display an error description in the Procedures tab. The modeler can then modify the program as needed and try to run it again.

3.3.4 Techniques for Running Simulation Experiments

Once a model is operating as intended, it can be used to run simulation experiments. There are several ways to do this. The most direct way to run experiments is for the user to actively manipulate the controls on the Interface tab to set up a simulation, then click the Run button and observe the model output.

The *BehaviorSpace* tool, which is accessed through the NetLogo Tools menu, provides a more sophisticated interface for experimenting with multiple related simulations in which parameter values are changed to compare alternate scenarios. It also provides a convenient workspace for modifying simulation setup details, stop conditions, or other aspects of the model for different experiments, including output reporting appearance and format. A powerful feature of BehaviorSpace is its ability to take advantage of multiple processors, if they are available on your computer, making it possible to run multiple simulation experiments concurrently.

Modelers who know how to write scripts can achieve even greater control over simulation experiments with BehaviorSpace by running it with Java from script files. The BehaviorSpace documentation provides easy-to-follow examples that demonstrate what kinds of tasks this approach is useful for and how to accomplish them.

Users who are proficient in Java can perform even more powerful simulations by writing Java code to create experiments that control the NetLogo model directly. The NetLogo Controlling Guide, part of the user's manual, provides substantial guidelines for manipulating NetLogo models using Java, including sample Java programs that can readily be adapted for this purpose.

3.3.5 Data Input and Output Capabilities

The NetLogo language contains a number of easy-to-use commands for opening, reading, and writing to data files. The sample model library contains a File Input Example and a File Output Example, and others, that demonstrate how input and output data files can be used in NetLogo models.

3.3.6 Extensibility

3.3.6.1 General Procedure

The power of NetLogo models may be significantly enhanced through the use of *extensions*, which are packages of code written in Java or other languages. Extensions provide model libraries of new commands, which effectively extend the NetLogo language with new capabilities. Extensions are added to individual models by reference, through the inclusion of the extension keyword in essentially the same way that external source code files can be incorporated into the basic model.

3.3.6.2 GIS Capabilities

NetLogo gains its GIS capabilities through a GIS extension that is bundled with the installation package. Like other extensions, the GIS extension becomes available to any model by including the GIS keyword in brackets after the extensions command at the top of the Procedures file, as shown below:

```
extensions [ gis ]
```

This extension gives NetLogo a useful set of GIS capabilities. Both vector (*.shp*) and raster (*.asc* and *.grd*) file formats are supported, as are a variety of projections. Once the desired GIS files are loaded and mapped to the NetLogo coordinate grid, a number of basic GIS functions become available, making it possible to link GIS data (spatial and attribute) with NetLogo's turtles and patches. As is typical, NetLogo documentation of these functions is exemplary. Also, the bundled GIS sample models provide informative, if brief, illustrations of many of these functions. Most of the case studies in this book illustrate the power that the NetLogo GIS extension can add to simple models.

3.3.6.3 Linking to External Software Packages

Analysis of data generated by NetLogo models has typically been accomplished by using external software such as spreadsheet, statistical, or mathematical applications to process data exported from the model. Recently, however, software applications developed by third parties have begun appearing to facilitate direct linking between NetLogo and external packages.

One product of interest is the NetLogo-R extension (developed by Jan C. Thiele and Volker Grimm, http://netlogo-r-ext.berlios.de), which provides automated links between NetLogo and the free statistical computing application, R (http://www.r-project.org/). This allows for runtime statistical analysis of NetLogo model output. Note that none of the examples in this book make use of this extension, but the R stats package was used to analyze results generated by the patch valuation model documented in Chap. 11.

Of interest to users of Wolfram Mathematica (Wolfram Research, Inc., Champaign, IL), the NetLogo installation package includes a plug-in called MathematicaLink. When installed in Mathematica, this add-on capability makes it possible to run NetLogo models from within Mathematica and use Mathematica capabilities to analyze and visualize a NetLogo model (Bakshy and Wilensky 2007).

3.3.6.4 Advanced Development of NetLogo Models

A new third-party tool is now available that greatly enhances the potential usefulness of NetLogo models. This tool, called ReLogo, provides an easy way to import NetLogo models into the Repast Simphony (Repast S) agent-based modeling platform for further development. Repast S, which was developed by Argonne National Laboratory for the U.S. Department of Energy (http://repast.sourceforge.net/repast_simphony.html), is a powerful Java-based platform for developing simulation models both in native Java code and the Java-based Groovy object-oriented programming language. ReLogo was integrated with Repast S 2.0 (beta) in 2010, and brings the benefits of the NetLogo coding language to the Repast modeler while providing access to the powerful features of the Repast environment.

ReLogo offers two very important prospective advantages to the greater simulation modeling community. First, it provides a powerful development path for simple NetLogo models that show potential for providing deep insight into poorly understood complex systems. ReLogo includes an import capability that translates NetLogo models for full functionality and development within Repast S. Second, ReLogo makes it easy for Repast S model builders to exploit the simplicity of the NetLogo environment for drafting quick working prototypes for the purpose of testing new ideas. Once a NetLogo-based prototype is validated and operational, and considered useful, it can then be migrated to Repast S for advanced development and integration into "enterprise-level models" (Ozik et al. 2007). Furthermore, the Repast platform includes a high-performance computing option that makes it possible to optimize Repast S models for execution on engineering workstations and cluster computers. Therefore, ReLogo offers the potential to capture simple models developed by subject matter experts who have little modeling capability and then transfer them to expert modelers capable of fleshing them out into powerful visualization, analytical, and planning tools.

3.4 Conclusions

As the models in this book demonstrate, NetLogo has been demonstrated many times over to be a viable platform for sophisticated model creation. It is a platform that is accessible to the subject matter expert who, regardless of the depth or subtlety of understanding of simulation modeling, may have little practical background as a hands-on programmer of automated models. NetLogo's ease of use, its well-designed interface, and its powerful and capable programming language are supported by outstanding documentation and a highly instructive collection of sample models, making it an outstanding tool for anyone who wishes to codify his or her expertise to explore the dynamics of a system.

References

Bakshy E, Wilensky U (2007) NetLogo-Mathematica link. Center for Connected Learning and Computer-Based Modeling, Northwestern University, Evanston. http://ccl.northwestern.edu/netlogo/mathematica.html. Accessed date 2/10/12

Ozik J, North MJ, Sallach DL, Panici JW (2007) ROAD map: transforming and extending repast with groovy. Proceedings of the agent 2007 conference on complex interaction and social emergence, Argonne National Laboratory, Argonne, Nov 2007

Wilensky U (1997) NetLogo wolf sheep predation model. Computer software. Center for Connected Learning and Computer-Based Modeling, Northwestern University, Evanston. http://ccl.northwestern.edu/netlogo/models/WolfSheepPredation. Accessed date 2/10/12

Wilensky U (1999) NetLogo. Computer software. Northwestern University, Center for Connected Learning and Computer-Based Modeling, Evanston. http://ccl.northwestern.edu/netlogo/. Accessed date 2/10/12

Chapter 4
A Simulation Model of Fire Ant Competition with Cave Crickets at Fort Hood, Texas*

Bart Rossmann, Tim Peterson, and John Drake

4.1 Background

Fort Hood, Texas, is home to rare endemic cave invertebrate species. Although these species are not listed under the Endangered Species Act (US Fish and Wildlife Service 1994, 2000), they are being carefully studied and managed to avoid listing. The at-risk invertebrates at Fort Hood are found in caves that extend across the entirety of the karst landscape, although many caves are concentrated in remote areas that are not often accessed by people.

The caves beneath Fort Hood lack primary producers and roosting bat populations, a large source of energy input in other cave systems. Therefore, the endangered karst invertebrates rely on organic matter vectored into the caves by crickets (*Ceuthophilus secretus*) (Taylor et al. 2003a).

Cricket guano, eggs, and juvenile nymphs are consumed by a number of gastropods (e.g., *Helicodiscus* spp. and *Mesodon* spp.), carabid beetles (*Rhadine reyesi*), and spiders (*Cicurina* spp.), respectively (Taylor et al. 2003a, b). Thus, cave crickets are a keystone resource supplier for the endemic karst communities.

These cave crickets also allow for a relatively easy method of measuring cave health, without a negative impact on the species of interest. While the endangered

*Based on ERDC/CERL TR-09-19, July 2009, with funding support from ERDC-CERL Project 140644, Habitat-Centric SAR (Species at Risk) Research—Multi-Species PVA (Population Viability Analysis).

B. Rossmann (✉)
Applied Technology for Learning in the Arts and Sciences, University of Illinois,
608 S Mathews, M/C 460, Urbana, IL 61801, USA
e-mail: bartmann@illinois.edu

T. Peterson • J. Drake
University of Illinois, Urbana, IL 61801, USA

J.D. Westervelt and G.L. Cohen (eds.), *Ecologist-Developed Spatially Explicit Dynamic Landscape Models*, Modeling Dynamic Systems,
DOI 10.1007/978-1-4614-1257-1_4, © Springer Science+Business Media, LLC 2012

species are hard to find, due to their relative lack of numbers and living deep within caves, *C. secretus* is populous and leaves the cave each night to forage.

However, the cave crickets are now at risk from an exotic species that has invaded Fort Hood: the Red Imported Fire Ant (RIFA) (*Solenopsis invicta*), which originated in South America and first appeared in the USA in the 1930s (Taber 2000).

The RIFAs spread quickly and successfully invaded numerous southern states. Among these are established colonies in Texas (Cokendolpher and Phillips 1989) that have been found within 15 m of caves in Fort Hood (Elliott 1992).

RIFA are aggressive omnivores that have been known to eat millipedes, salamanders, earthworms, and both live and dead cave crickets (Elliott 1992; Wojcik et al. 2001). This makes them a broad spectrum pest to the management of any endangered invertebrate species within their territory. A number of observational studies have documented the potential for detrimental impacts of RIFA on the karst community. Taylor et al. (2003b) documented widespread colonization of Fort Hood by RIFA, including those areas with caves.

Both cave crickets and RIFA are opportunistic omnivores that eat nearly everything except raw plant materials (Campbell 1976; Elliott 1992; Taber 2000; Cokendolpher et al. 2001). Thus, interspecific competition for resources is likely. In addition, RIFA foraging of cricket eggs and nymphs has been observed inside caves on Fort Hood (Reddell 2001; Taylor et al. 2003a), especially in summer months when high temperatures drive RIFA deep into the soil. Elliott (1992) has observed RIFA preying upon live adult crickets, along with many other cave invertebrates.

These field studies have motivated the management of fire ant mounds in the vicinity of caves, in order to protect karst invertebrates. However, no model to date has investigated the broader impacts of a different management strategy as it affects cricket populations and, ultimately, the sustainability of karst communities and fire ant abundances on a landscape scale. We have developed an expedient simulation model that provides the ability to track RIFA agents in the form of mounds across a spatial and temporal landscape, and to assess the impact of the invasive species on the native community.

4.2 Objectives

The objectives of this model are to incorporate the information from field studies into a spatially explicit model of fire ant and cricket behavior, and to document the effectiveness of RIFA management in order to ensure the long-term sustainability of karst communities. We hypothesize that the size of cave cricket populations, eradication of RIFA mounds, and intensity of RIFA within the foraging area of cave crickets affect the probability of the destruction of the cave invertebrate populations. We expected that the results would help to identify the level of management needed for caves of varying size, as well as direct areas of future research.

A general objective of this work was to demonstrate how biologists and land managers may quickly develop a simple computer-based model, using location-specific data and parameters with public-domain software, that can rapidly simulate the impacts of alternate habitat-management strategies. The intent is to illustrate that simple, expedient models may be developed by personnel who have no expertise in model building, and how those models may add considerable value to land management activities.

The RIFA model documented here was created for use in the public domain, and it may be downloaded from http://earth.cecer.army.mil/LandSimModel/?q=node/43 at no charge for adoption or modification by natural resource managers working in RIFA territories. Although this model was developed for karst environments at Fort Hood, it can readily be modified to address RIFA management questions at other locations.

4.3   Model Description

4.3.1   Purpose

The specific purpose of the model was to understand the potential impact of RIFA foraging and raiding on native karst fauna (cave crickets), particularly if fire ant populations continue to grow at rates consistent with the last 50 years. Two goals for the model were to provide a tool to help in the assessment of RIFA management and to evaluate potential parameters of interest for future field research. Spot mound eradication using hot water is a common management technique and will be tested in the model over a period of 10 years. The model will also assess various parameters (cave-carrying capacity, intrinsic rate of growth, cave raiding by RIFA) to understand which parameters cause significant deviance in cave viability, both with and without management. This will focus further studies in data acquisition for cave managers, and provide a framework for cost-effectiveness of future RIFA management.

4.3.2   State Variables and Scales

Interactions between RIFA, cave crickets, and resources in the environment were modeled using NetLogo 4.0.2 (Wilensky 1999).[1] RIFA and crickets are modeled as colonies, labeled "Mound" or "Cave," respectively. Both obtain resources from the environment, but only mound colonies reproduce. Resources obtained from the environment directly influence the number of individuals contained within the mound or cave. An overview of interactions within the model is shown in Fig. 4.1.

---

[1] An operational copy of this model is available through http://extras.springer.com.

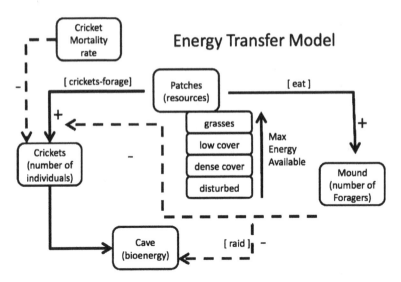

Fig. 4.1 Overview of interactions within model

4.3.3 Design Concepts

4.3.3.1 Colony Invasion

RIFA mounds, or colonies, consist of a queen and an array of supporters that carry out the various functions of the colony, including foraging, construction, and defense. These colonies act as a single organism that reproduces by spawning new colonies in new mounds nearby. This process is captured in the model as a seasonal activity.

4.3.3.2 Predator–Prey

We modeled caves and RIFA mounds, rather than individual crickets and RIFA, as agents. This approach dramatically improved modeling speed by reducing the number of agents from many thousands to several dozen. Mounds were associated with a number of RIFA, and caves with a number of crickets. Each agent was associated with a foraging distance and where cricket and RIFA foraging overlapped, and predator–prey relationships determined changes in cricket populations in caves and the associated energy changes to both mounds and caves.

4.3.3.3 Energy Networks

Figure 4.1 captures the energy exchange network. Crickets gain energy from foraging on the landscape around caves. RIFA forage in the same area but also prey on the cricket population.

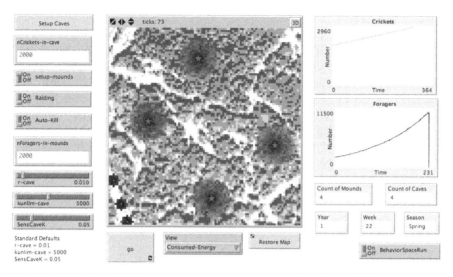

Fig. 4.2 World view of NetLogo model area shown as imported and coded LCID image. *Dark gray* disturbed area (e.g., dirt roads, land development); *white* grassland; *black* low density cover (e.g., shrubs and small trees); *light gray* high density cover (e.g., high density trees)

4.3.4   Initialization and Input

4.3.4.1   World Details

Satellite map images were downloaded from US Geological Survey (USGS) Seamless Server (http://seamless.usgs.gov), and coded with NLCD 2001 Land Cover Data.

ERDAS IMAGINE (http://www.erdas.com/) was used to convert each pixel into one of four colors, each matched to a specific type of land cover. Upon importing the image into the NetLogo model, the colors were changed for ease of reference into the following:

- DARK GRAY = disturbed area (e.g., dirt roads, land development).
- WHITE = grassland.
- BLACK = low density cover (e.g., shrubs and small trees).
- LIGHT GRAY = high density cover (e.g., high density trees).

Each pixel coded represents a 10×10 m plot of land. The finished world view can be seen in Fig. 4.2. Within the model's world, there are intermittent areas of high and low cover, along with diagonal streaks of disturbance that may have resulted from vehicle movement, the collapse of trees, or human activity. In the lower left corner of the model's world is a streak of disturbance. This is the starting point for the introduction of mounds. From here, three mounds are introduced and will propagate across the landscape test location as detailed below.

4.3.4.2 Resources

Each imported pixel is represented in NetLogo by a *patch* and assigned a maximum-energy variable based on the specific land type. To simulate replenishment of resources from influx of prey species, growth of edible plant matter, etc., the available energy of each patch has a 20% chance of increasing during each week, up to its maximum-energy variable. This prevents the caves and mounds from depleting their resources and dying off, and creates a more realistic setting of use and growth of resources. Caves and mounds will attain a stable carrying capacity based on the local resources available, and the resource replenishment within their respective foraging range.

4.3.5 Submodels

4.3.5.1 Cave Crickets

Cave stabilization. Within the model, the caves are given a period of time without competing with mounds, or being preyed upon by mounds. This is to allow for each cave to reach a stable population size comparable to that found in the wild. It takes approximately 5–10 years for the cricket population to stabilize, without any fire ants present. (The model allows for 10 years to pass before the introduction of mounds).

Foraging range. Caves rely on foraging crickets to bring in bioenergy. The minimum and maximum range for crickets' foraging distance from the cave is 30 and 100 m, respectively (Taylor et al. 2003a). Caves with larger populations of crickets have a greater range. Below 30 m from the cave, cricket density during foraging is uniform. At greater ranges (up to 100 m), the density of crickets drops, ultimately reaching a density of no crickets at 100 m.

Since the movements of individual crickets are not tracked in this model, this function of cricket density allows for the reduction of available energy to the crickets—the further the patch from the cave, the less energy will be gathered, due to lower cricket density. This simulates both the decreased number of crickets at distant ranges from caves as well as the reduced energy consumption at the further ranges, due to more energy being expended while retrieving nutrients farther from the cave.

Birth and death. Cricket populations were determined using the Verhulst equation, with varying carrying capacity used to simulate the different sizes of caves at Fort Hood. The intrinsic rate of growth was determined to be within the range of closely related cricket species, and below that of the faster-reproducing RIFA.

The equation also included a sensitivity to the surrounding conditions, where the impact of decreases in available energy could vary. Within the formula used, a lower number indicated a high sensitivity to available resources.

4.3.5.2 RIFA Model

Mound foraging range. The RIFA mound sub-model's foraging code is virtually identical to the cave sub-model, due to the similar energy flow. For a RIFA mound, a large percentage of the adult ants are foragers, and leave the mound to gather energy. Their success in bringing in energy means the population of the mound increases, reaching a maximum-carrying capacity of 250,000 ants (Markin et al. 1974). Since RIFA are observed to be more aggressive, more cooperative and more numerous in their foraging than crickets, RIFA will usually gain more energy from the same source when mounds and caves simultaneously forage the same patch of resources.

Mounds will also diminish the available resources in a patch during each week before the caves are able to forage. This is also due to the greater aggression of RIFA over the crickets, as areas that have RIFA show a decrease in crickets numbers (Taylor et al. 2003a). It is believed that the presence of RIFA discourages cricket foraging as they attempt to avoid predation. RIFA also forage during both day and night, but are more dependent on temperature than time of the day. By contrast, cave crickets forage only at night.

Mound propagation. RIFA typically propagate during mean daily temperatures over 69°F but below 89°F (Tana 2002), and with high humidity or rain present. By mapping the weekly mean temperatures together with average rainfall and humidity, we can designate propagation seasons for RIFA (spring and fall in this model). The propagation seasons cause the spread of RIFA across the landscape and higher foraging activity, since late propagation requires enormous energy. Propagation is a high-risk activity for RIFA; only 2% of new starts typically result in a successful new mound.

Three factors affect the success of mound establishment: (1) new mounds participating in intraspecific mound raids, (2) new mounds having enough energy resources nearby without high competition, and (3) new mounds being more successful in disturbed areas.

To simulate these conditions within the model, new mounds followed three sets of rules: (1) If multiple new mounds are established on the same patch, they will form into one mound, simulating mounds performing raids in which the losing mounds' workers are absorbed by the winning mounds. (2) New mounds will not be able to establish on the same plot as old mounds. (3) New mounds will be more likely to establish on disturbed areas.

Mound raiding. Caves at Fort Hood are periodically raided by RIFA. This typically occurs in the summer months when RIFA stay underground to avoid the heat (Taylor et al. 2003a). To simulate this within the model, caves within the foraging range of a mound were given a 20% chance of being raided each week during the summer season.

Also within the model, caves that are raided lose 100 crickets. This number was chosen as a conservative estimate of the raiding that typically occurs close to the entrance of a cave, or the Twilight Zone, though it is possible that additional raiding occurs in remote regions not accessible by humans (Taylor et al. 2003a).

Mound management. One common method employed to exterminate RIFA is the injection of boiling water deep into the mounds. Within the model, management of mounds simulated a yearly hot water treatment of RIFA mounds within the foraging radius of caves.

When enabled, each year mounds within the foraging radius of a cave would have a 60% chance of being killed. This is because injecting hot water into the mounds has a 60% success rate, which follows observed data (Nature Conservancy 2000), and allows for reduced, though still present, impact on the caves by mounds.

4.4 Simulation Experiments

Results were analyzed using the general linear model software from the SAS company (http://www.sas.com). Because the random effect for caves was not significant, no random effect was fit to the model. The Tukey–Kramer procedure for analyzing unequal pair-like comparisons was used to adjust for multiple comparisons. No outliers were present, and all runs under all parameters were used.

Most main effect and combined effect parameters within the model were statistically significant. This shows that the presence of RIFA significantly impacts cricket populations at all cave levels ($p = 1.000$). The parameters are also useful in understanding that RIFA significantly impacted crickets, whether or not raiding was turned on, as this could possibly be a point of contention (see Sect. 4.5).

However, when too much data are considered significant, it runs the risk of being meaningless. We mitigated this risk with two further steps. The first step was to account for what was not statistically significant. Establishing those parameters allowed us to understand when no change needs to occur in order to protect cricket populations between situations. These parameters also allowed us to check for model reliability, by comparing situations that should not be significantly different. The second step was to identify select cases of importance, such as complete cave loss, and large patterns that are not illuminated by the statistical analysis. The following paragraphs describe these steps in detail.

4.4.1 Model Validation

To check for model reliability, we examined the situations where management techniques were applied, both with and without RIFA present. With no RIFA present, management techniques should not cause any change in the average number of crickets. This was confirmed by our results: average number without RIFA, management on = 91.8%; management off = 91.8%; $p = 1.000$.

However, all other conditions with RIFA being present, whether raiding was turned on or off, and whether management was turned on or off, were highly significant from

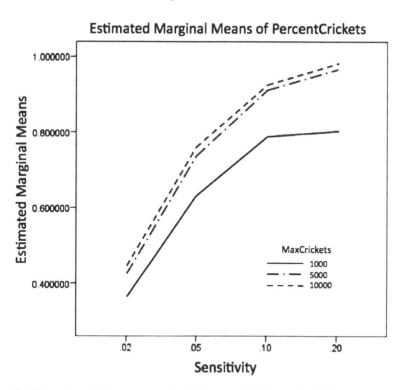

Fig. 4.3 Effect of sensitivity on average level of percent of crickets when RIFA is present

each other ($p<0.0001$). This is evidence that management techniques within the model were affecting cricket populations through reduction of RIFA, and not an unknown error in the coding.

Within any setting of sensitivity to resources, the caves with a $K=1,000$ were consistently at a disadvantage to caves with a $K=5,000–10,000$ (Fig. 4.3).

Caves with a $K=5,000–10,000$ did not have statistically significant differences in their average population across all levels of sensitivity ($p=0.9533, 0.7179, 0.2905$, and 0.3259 for sensitivity; $0.1, 0.2, 0.02$, and 0.05 for respectively). However, caves with a $K=1,000$ crickets saw significantly fewer percentage of those crickets survive, though the general trend followed that of the caves with a higher K value. Sensitivity appears not to affect caves severely when they have a $K=5,000$ or higher, but caves with a $K=1,000$ appear to be impacted greatly by sensitivity.

Concerning the overall average number of crickets, caves that had a $K=1,000$ were significantly different from caves that had $5,000–10,000$ crickets ($p<0.0001$). While caves that had $5,000–10,000$ crickets were significantly different from each other ($p=0.0002$), the averages were close enough (81.4 and 82.4 %, respectively) that this significance can be attributed more to the large sample size than to an effect between caves that needs to be accounted for with management procedures.

At the most robust sensitivity level (0.2), the presence of RIFA does not significantly affect the number of crickets when they are not raiding caves ($p = 0.9205$). Nor is there a significant change from these if the sensitivity is dropped to 0.1, so long as RIFA are no longer present ($p = 1.000$ when compared to sensitivity 0.2 and no RIFA present, and $p = 0.9899$ when compared to sensitivity 0.2 and RIFA present, but not raiding caves).

Raiding caves plays a role in differentiating the large and small caves. If RIFA are present and raiding, they impact cave populations at all cave sizes ($p = 1.000$); however, there is no difference between how they impact cave sizes when raiding is turned off ($p = 1.000$). Once raiding is turned on, the number of cave crickets significantly drop at all cave sizes ($p < 0.0001$).

Cave sizes have been seen to ameliorate some of the effects of RIFA. Management significantly increases the level of crickets at all cave sizes ($p < 0.0001$). However, in some cases the same result can be seen through an increase in cave size. To elaborate, the average number of crickets in the smallest cave ($k = 1,000$), when management is turned on, is not significantly different than the next largest cave ($k = 5,000$), when the larger cave does not have management ($p = 1.000$).

Management can also be seen to decrease the number of crickets lost in two larger caves. When management is applied, there is no significant difference between caves of $K = 5,0000$ or $K = 10,000$ ($p = 0.5286$). Without management, these two cave sizes are significantly different ($p = 0.0004$), though not as much as is normally seen. As their average population sizes are decreased by less than 2% (80.5 and 79%, for the 10,000 and 5,000 K caves, respectively), this can again be attributed to the large number of runs rather than a decrease that warrants concern.

4.4.2 Hypothesis Testing

If no RIFA were introduced, caves continued to stay at, or close to, their K value. However, if mounds were introduced, caves showed a decrease in the number of crickets they held. This decrease led to the loss of entire caves in nine separate conditions, as shown in Table 4.1. Several other conditions showed a severe decrease in cricket population, but those decreases did not result in complete population loss in any of the four caves in each simulation.

Table 4.1 is sorted by the average number of caves that lost all crickets. The average numbers of caves (out of four in each simulation) that "died" are listed, along with the standard deviation for each design. With one exception, each loss happened to a cave with the lowest maximum carrying capacity. The exception had the next-highest carrying capacity, and the maximum level of sensitivity to resources, with ants present and raiding, and no hot water management of the ants. All possible designs that included small caves with raiding ants experienced some cave loss. This result suggests that smaller caves are more at risk to species loss than caves with large carrying capacity, particularly if the ants use the caves directly for resources as opposed to only competing for outside foods. It also shows that a moderate level of management for the ants may not be sufficient to prevent species loss in smaller caves, as management reduced but did not eliminate complete cricket loss from smaller caves.

Table 4.1 Conditions leading to loss of all crickets

Sensitivity	Max crickets	Raiding	Management	Ants	AveDead	StDevDead
0.02	1,000	TRUE	FALSE	TRUE	2.323	0.979
0.05	1,000	TRUE	FALSE	TRUE	1.839	1.003
0.2	1,000	TRUE	FALSE	TRUE	1.821	0.983
0.1	1,000	TRUE	FALSE	TRUE	1.643	1.096
0.02	1,000	TRUE	TRUE	TRUE	0.774	0.845
0.2	1,000	TRUE	TRUE	TRUE	0.143	0.356
0.02	5,000	TRUE	FALSE	TRUE	0.097	0.301
0.1	1,000	TRUE	TRUE	TRUE	0.071	0.262
0.05	1,000	TRUE	TRUE	TRUE	0.065	0.250

Of the nine conditions that lost the most caves, eight had a $K = 1,000$ crickets. This accounted for every condition that had a $K = 1,000$ crickets, with ants present, and raiding turned on. Sensitivity and management implementation had an effect on the average number of caves that lost all crickets—with sensitivity effects having priority over management effects. But, all conditions with $K = 1,000$, and ants performing raids, had at least some where there was a total loss of crickets.

Caves showed a marked reduction in losses when management was turned on, from 1.643 to 0.774. This is despite the finding that the condition having the "worst" sensitivity setting did better, while the condition having the "best" sensitivity setting for caves lost over 1.5 caves, on average.

The only condition to experience complete cricket loss, other than the above-mentioned trend of caves with a $K = 1,000$, was a single condition that included a $K = 5,000$. This condition also included the "worst" case scenarios for the caves: high sensitivity, ants present, raiding turned on, and management turned off. However, this condition only lost, on average, 0.097 caves out of four. Likewise, the average number of crickets in caves under this condition was 1,801. Also, no other conditions that included a $K = 5,000$ lost any crickets.

Similarly, no caves with a $K = 10,000$ crickets experienced complete loss of crickets. However, the lowest number of crickets for a cave with a $K = 10,000$ was 4,045. This is a reduction of over half the maximum crickets, though not a reduction that places the cave at a severe risk of cricket loss.

Large reductions are not uncommon for larger caves, though. While the smaller caves were the most likely to have lost all crickets, caves with a $K = 10,000$ made up a significant portion of those caves that had a severe reduction in the percent of crickets remaining at the end of the trials (shown in Table 4.1). Larger cave populations are still highly impacted by RIFA activity, with several populations losing over half of their maximum capacity.

However, these larger caves stabilize at those reduced populations, whereas the smaller caves cannot support such large reductions. While the largest caves lost over half their populations in some cases, none was completely wiped out. Bottlenecking of the population gene pool may be a problem for larger caves, but loss of population is not as much a problem for the largest caves as it is for the smallest caves.

Table 4.2 Simulation results sorted by the percent of surviving crickets

Sensitivity	Max crickets	PercentMax	Raiding	Management	Ants
0.02	1,000	0.186314 52	TRUE	FALSE	TRUE
0.05	1,000	0.32076613	TRUE	FALSE	TRUE
0.02	1,000	0.33295161	TRUE	TRUE	TRUE
0.02	5,000	0.3602129	TRUE	FALSE	TRUE
0.02	10,000	0.40452903	TRUE	FALSE	TRUE
0.2	1,000	0.41138393	TRUE	FALSE	TRUE
0.02	5,000	0.43519516	FALSE	FALSE	TRUE
0.02	10,000	0.43715	FALSE	FALSE	TRUE
0.02	1,000	0.43952419	FALSE	FALSE	TRUE
0.1	1,000	0.4429375	TRUE	FALSE	TRUE
0.02	5,000	0.44355484	TRUE	TRUE	TRUE
0.02	10,000	0.47659435	TRUE	TRUE	TRUE
0.02	1,000	0.48024194	FALSE	TRUE	TRUE
0.02	10,000	0.48645806	FALSE	TRUE	TRUE
0.02	5,000	0.48657097	FALSE	TRUE	TRUE
0.05	1,000	0.65820161	TRUE	TRUE	TRUE
0.05	5,000	0.66205968	TRUE	FALSE	TRUE
0.05	10,000	0.71096129	TRUE	FALSE	TRUE
0.02	1,000	0.71609375	FALSE	FALSE	FALSE
0.02	5,000	0.71629375	TRUE	FALSE	FALSE
0.02	5,000	0.71637188	FALSE	TRUE	FALSE
0.02	1,000	0.71645313	TRUE	FALSE	FALSE
0.02	5,000	0.7164625	FALSE	FALSE	FALSE
0.02	10,000	0.71648047	TRUE	TRUE	FALSE
0.02	10,000	0.71656875	FALSE	FALSE	FALSE
0.02	10,000	0.71663203	TRUE	FALSE	FALSE
0.02	5,000	0.71667188	TRUE	TRUE	FALSE
0.02	10,000	0.71667656	FALSE	TRUE	FALSE
0.02	1,000	0.716875	FALSE	TRUE	FALSE
0.02	1,000	0.71692188	TRUE	TRUE	FALSE
0.05	5,000	0.73736452	FALSE	FALSE	TRUE
0.05	10,000	0.75688917	FALSE	FALSE	TRUE
0.05	1,000	0.75910484	FALSE	FALSE	TRUE
0.05	5,000	0.77403065	TRUE	TRUE	TRUE
0.05	10,000	0.78235484	TRUE	TRUE	TRUE
0.05	1,000	0.79287097	FALSE	TRUE	TRUE
0.05	5,000	0.79536452	FALSE	TRUE	TRUE
0.05	10,000	0.79562581	FALSE	TRUE	TRUE

While reduction relative to K is widespread across all levels of K, those with the fewest absolute number of average crickets are the caves with a $K = 1,000$.

Surprisingly, though, the correlation between the average cricket numbers and the number of caves lost is only -0.504 (correlation only for the 35 smallest caves). The average number of crickets is not tightly proportional to the number of caves lost, and there are even several conditions with lower average cricket numbers and no caves lost than some conditions with caves lost (Table 4.2).

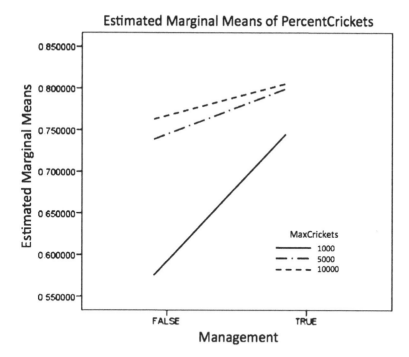

Fig. 4.4 Effect of management on average level of percent of crickets when RIFA is present

The average number of crickets indicates a cave that is at risk of complete cricket loss, but not definitively. At 500 or fewer average crickets, there is likely to be a loss of at least one cave. At average cricket populations of 2,000 or fewer, cave loss is still possible, though far less likely.

While the presence or absence of RIFA is the main contributing factor to the overall percentage of crickets that survive, the sensitivity to resources plays a larger role than either the presence or absence of hot water management of the RIFA mounds, RIFA's ability to raid caves, or the absolute carrying capacity of the cave. For caves within the foraging radius of RIFA, the robustness of cricket populations to fluctuations of resource availability is a key factor that requires further data collection and analysis.

At any cave size, management of mounds was able to alleviate the decrease in crickets caused by RIFA activity. However, for caves of $K = 1,000$, management was seen to play a much larger role, with the average population size increasing by almost 20% of its maximum (see Fig. 4.4).

While larger caves do appear to be influenced by yearly hot water treatment of surrounding RIFA mounds, this indicates that smaller caves may be the most cost-effective place to provide RIFA management. This is especially true since the lower average number of crickets present in caves (where $K = 1,000$ when no management is present), is due in part to the relatively large number of caves that have lost all their crickets when no management was applied.

With management, the largest average number of caves lost was 0.774, but without management, the largest average number lost was 2.323, as was shown in Table 4.1.

4.5 Discussion

RIFA are known to be a danger to cave communities (Taylor et al. 2003b). With the RIFA invasion of Fort Hood, Texas (Elliott 1992), cave-dwelling species listed under the Endangered Species Act (USFWS 1994, 2000) are being threatened. The cave cricket simulation model shows that not all caves will be impacted uniformly by RIFA, but additional information about the robustness of cricket populations is needed to understand the scope of impact.

In the simulation model, five factors influenced cave cricket survivorship:

1. Presence or absence of RIFA.
2. Whether or not RIFA raided caves.
3. K of the cave, which can be correlated in natural populations to overall cave size and abundance of surrounding resources.
4. Sensitivity of cricket populations to fluctuations in resource availability.
5. Presence or absence of the hot water treatment performed on caves within the foraging radius of crickets.

The model confirmed that RIFA can have a negative impact on cave cricket populations. While this impact varies in severity across many scenarios, it caused complete cave loss a significant number of times. Even without complete cave loss, the sharp decline in the population of cave crickets causes a reduction in their gene pool. Even the largest caves examined sometimes lost more than half their cricket populations. Because the model showed that RIFA can seriously impact cave cricket populations even without raiding activity, RIFA management is needed even when raiding is found to be moderate compared with expectations.

However, in a significant number of simulations, raiding did play an important role in complete cave loss. As noted above, a conservative estimate was assumed for cave cricket loss to RIFA during raiding because the exact number is not known due to difficulties in tracking RIFA raids. If that estimate is too low, then raiding could account for a significant portion of the RIFA problem, and management techniques that specifically target raiding would have to be devised. The most important point to remember is that the model shows RIFA impacting cave communities even when raiding ceases.

The RIFA model showed that cave size is a major factor in the loss of caves. Despite caves of all sizes having their average percent of the population lowered, the caves with a $K = 1,000$ crickets were the most at risk for complete loss. This may occur through the lowering of the absolute cricket populations, at which point, either raiding or lack of resources may be the final stress that removes all crickets.

For purposes of predicting total cave loss, the number of crickets within a cave (once it has stabilized to a RIFA invasion) is far more important than the number of

Table 4.3 Simulation results sorted by the average number of crickets left at the end of the trial

Sensitivity	Max crickets	Ave crickets	Ave cave lost	Percent of max
0.02	1,000	186.3145161	2.322580645	0.186314516
0.05	1,000	320.766129	1.838709677	0.320766129
0.02	1,000	332.9516129	0.774193548	0.332951613
0.2	1,000	411.3839286	1.821428571	0.411383929
0.02	1,000	439.5241935	0	0.439524194
0.1	1,000	442.9375	1.642857143	0.4429375
0.02	1,000	480.2419355	0	0.480241935
0.05	1,000	658.2016129	0.064516129	0.658201613
0.02	1,000	716.09375	0	0.71609375
0.02	1,000	716.453125	0	0.716453125
0.02	1,000	716.875	0	0.716875
0.02	1,000	716.921875	0	0.716921875
0.05	1,000	759.1048387	0	0.759104839
0.05	1,000	792.8709677	0	0.792870968
0.2	1,000	826.6696429	0.142857143	0.826669643
0.1	1,000	857.9196429	0.071428571	0.857919643
0.1	1,000	918.2678571	0	0.918267857
0.1	1,000	950.4375	0	0.9504375
0.05	1,000	957.1875	0	0.9571875
0.05	1,000	957.25	0	0.95725
0.05	1,000	957.2578125	0	0.957257813
0.05	1,000	957.265625	0	0.957265625
0.2	1,000	987.6160714	0	0.987616071
0.2	1,000	995.4732143	0	0.995473214
0.1	1,000	998	0	0.998
0.1	1,000	998	0	0.998
0.1	1,000	998	0	0.998
0.1	1,000	998	0	0.998
0.2	1,000	1,000	0	1
0.2	1,000	1,000	0	1
0.2	1,000	1,000	0	1
0.2	1,000	1,000	0	1
0.02	5,000	1801.064516	0.096774194	0.360212903
0.02	5,000	2175.975806	0	0.435195161
0.02	5,000	2217.774194	0	0.443554839

crickets the cave can hold. However, smaller caves are still important; the 32 caves with the lowest number of average crickets all had a $K=1,000$, while there was a condition in which caves starting at $K=5,000$ experienced significant loss.

However, the average number of crickets left at the end of a simulation was not a direct indicator of cave loss. The eight conditions in which the caves had a $K=1,000$ and experienced cave loss were not the eight caves with the lowest average number of crickets (Table 4.3). These caves, and their relative number of complete losses, suggest that the caves be narrowed down to two groups: those that may experience cave loss and those that are at high risk for cave loss.

While complete cricket loss is slightly correlated to average number of crickets ($r=-0.504$, indicating that as the average number of crickets increases the likelihood of cave loss decreases), there are many conditions with no cave loss that result in fewer average crickets than some conditions with some cave loss. A lower average number of crickets puts caves at risk for complete cricket loss, but is not an absolute indicator that caves will be lost.

Caves with a population averaging below 500, when RIFA are present, can be considered at high risk for cave loss. Caves under these conditions often experienced complete loss of crickets, with over five out of seven conditions resulting in complete loss of crickets. This loss occurs regardless of the sensitivity crickets have to their resources, or if RIFA are directly raiding the caves, or whether management techniques are in place (although management can be seen to reduce the number of caves lost).

Conditions where the caves have an average population of more than 500 crickets still show same caves loss, though it drops off drastically, with 0.14 average caves lost being the largest number of caves that lose all crickets. However, there are still possible losses of caves all the way to those having an average population of 2,000 crickets. Despite a relatively robust level of survivorship, these caves also appear to be at risk of cricket loss from RIFA.

While RIFA also appeared to impact the caves that have larger cricket populations, in some cases lowering their populations to 40% of their K, those caves appeared to settle at a lower new population without risk of complete cricket loss. Therefore, efforts to reduce RIFA foraging close to those caves may not be cost-effective for conserving the endangered species within the cave ecosystems on Fort Hood.

Conversely, smaller caves appear to be impacted greatly and are in need of at least yearly hot water treatments to surrounding RIFA mounds. In many cases, even this treatment may not be enough, and more aggressive management measures may be needed.

Sensitivity to surrounding resources was also a factor in the survivorship of cave cricket populations. Unfortunately, it is not known how sensitive crickets are to fluctuations in resource availability, and this is not a factor that can be controlled for as can controlling for population size. Sufficient information is not available to understand how cricket populations fluctuate with resource availability and the degree to which resource availability is changed by the addition of RIFA. Until this is known, management techniques need to be aggressively applied to all Fort Hood caves with small cricket populations, and at least yearly spot treatment performed for caves with moderate cricket populations.

The results of the cave cricket simulations suggest two courses of action. First, they point to the need for field research on how cave crickets respond to changes in food availability. This information is needed because it helps to determine which caves need to be protected from RIFA, and how aggressively. Depending on cricket sensitivity to food availability, it may be found that some caves need no protection while additional caves need to be managed. Second, the model shows us potential impacts of RIFA on cave communities, operating through their impact on cave crickets, which demands immediate management to prevent species loss.

The simulations have shown that a simple hot water treatment can effectively reduce loss of species in most caves and also that more aggressive treatment is needed for smaller caves until more is known about how cricket populations react to changes in resources. At that time, we may find that hot water treatment is still effective or that more management is needed to protect endangered karst invertebrates.

This discussion highlights some of the limitations of out RIFA model. While some parameters were accounted for through multiple iterations that assigned different values (sensitivity to resources being one of these), others had to be estimated to simplify the model enough that it could be run. Among these, and related to sensitivity, is the question of how quickly resources are depleted by each species.

Resource limitations can be assumed to play a role in regulating population size, but it is not known how resources are depleted. Neither is it known how fast resources are replenished through new growth, the influx of new prey species, or other means. It is also likely that the rate of replenishment will change on a seasonal basis. Future versions of the RIFA model should account for various rates of growth during each season.

The overall populations of RIFA mounds may vary with the season. There has been evidence that RIFA populations reach their maximum numbers in midwinter, their maximum biomass in the spring, and declined to a minimum population in midsummer (Tschinkel 1993). However, it can be difficult to track exact numbers of RIFA if they are foraging more underground to avoid the summer heat, or if they are part of polygyne communities with multiple queens sharing control over a single colony with multiple mounds.

To simplify control methods, this model incorporated the most common method for exterminating RIFA, i.e., hot water treatment. Other options are available, however, including pesticides, poison bait and the imported phorid fly. While each of these options poses its own risk, all have been used to some degree, and it would be beneficial to understand the impacts of these methods on karst fauna.

One problem with all management interventions is the method of application. Most areas needing management are not easily accessible by humans. Bringing in equipment to control RIFA carries the risk of creating habitat disturbances that can actually facilitate RIFA colonization. For example, conveying boiling water to RIFA mounds requires a truck with a large enough bed to hold a boiler, hoses, and ancillary equipment. Driving to a treatment area and around each mound can uproot vegetation and soil, providing new routes for RIFA to more easily access the site and potentially colonize the entire area around the cave. Obviously, then, hot water treatment equipment can potentially be a very counterproductive way to approach the preservation of endangered kvarst invertebrates. To better understand the disturbance mechanisms and impacts of applying hot water to mounds, future versions of the RIFA model will include the capability of creating disturbances at a set point in relation to the caves and surrounding region during each treatment session.

4.6   Conclusions and Recommendations

4.6.1   Conclusions

This chapter describes the development of a simple computer simulation model that has served us as a cost-effective decision support tool for the proactive management of karst environments hosting species at risk of predation by RIFA. This model captures the expert knowledge of natural resources personnel and combines it with data from field studies and GIS information to create a spatially explicit model of RIFA behavior and its impact on cave cricket populations. The model was developed using the public domain NetLogo modeling environment, and did not require the intervention of a computer programmer. Ecologists and biologists need no computer expertise to develop NetLogo models, and the results are transparent enough to be understood by other technical peers with no computer expertise.

While providing valuable, actionable insights into karst population vulnerability to RIFA predation as related to cricket population sizes, cave sizes, and RIFA activities, the model is currently limited by several gaps in knowledge about the subject ecosystem. Three key gaps are as follows:

1. The impact of RIFA invasions on resource availability to cricket populations. Filling this gap would require field research to document the amount of pertinent resources depleted from the environment over a set time and how fast they are replenished, as well as the degree to which the presence of RIFA discourages cave crickets from foraging in the same area.
2. The impact of food variation on cricket populations. In order to better understand this dynamic, field research is needed to quantify the relation between a given decrease in food availability and the resulting decrease in cave cricket population. Combined with new field data about RIFA impact on resource availability, this information will provide a basis for the simulation model to quantify the indirect impact of RIFA on cave crickets.
3. The extent of RIFA raiding in Fort Hood caves. Field documentation would be difficult because many cave entrances are too small for direct human access. Although current simulations indicate that raiding is not a strong factor in decreasing the average number of crickets in a cave, raiding did play a strong role where entire cricket populations were lost. One possible interpretation is that RIFA raiding may provide the "final blow" in eliminating cricket populations that have been sufficiently stressed through other mechanisms, but field research would be necessary to supply data that the model needs to test that hypothesis.

These identified gaps pertain specifically to the Fort Hood cave environment. They may or may not represent information gaps that would be relevant to enhancing the RIFA model for application at other locations. The general point is that the NetLogo cave cricket model may be extended with additional site-specific data to improve the realism of its simulations.

4.6.2   Recommendations

Based on our interpretation of simulation results, we were able to offer the following recommendations about RIFA management to Fort Hood personnel.

1. Cave ecosystems that rely on cave cricket populations need to be identified and assessed. The locations of these caves and the types of surrounding environment (e.g., disturbed or not, relative abundance of resources) need to be recorded, along with recording the population of crickets within the cave, and the presence or absence of RIFA.
2. Caves that have RIFA present will need management at different levels, depending on the cricket population present.
3. Caves with a population of 500 or fewer crickets need immediate protection. All surrounding RIFA mounds should be treated with hot water, and treatment should continue regularly on any new RIFA mounds that appear. The caves should be monitored to assess the health of the cricket population.
4. Caves with a population of 2,000 crickets and the presence of RIFA should have their RIFA mounds treated with hot water on a yearly basis, at minimum.
5. Caves with a population of 1,000 crickets, but that do not yet have RIFA present, need to be monitored closely to see if any RIFA invade the habitat. If RIFA invade, they need to be managed on a yearly basis while watching cricket populations to see if more management is required.
6. Caves that have a population of 5,000 crickets, with no RIFA present, should be inspected on a yearly basis to see if RIFA invade, and, if so, how the population is affected.

References

Campbell GD (1976) Activity rhythm in the cave cricket, *Ceuthophilus conicaudus* Hubbell. American Midland Naturalist 96(2):350–366

Cokendolpher JC, Phillips SA Jr (1989) Rate of spread of the red imported fire ant, *Solenopsis invicta* (Hymenoptera: Formicidae), in Texas. Southwest Nat 34(3):443–449

Cokendolpher JC, Lawrence RK, Polyak VJ (2001) Seasonal and site-specific bait preferences of crickets and diplurans in Hidden Cave, New Mexico. Texas Memorial Museum, Speleological Monographs, vol 5. University of Texas, Austin, pp 95–104

Elliott WR (1992) RIFA and endangered cave invertebrates: a control and ecological study. Final report submitted to Endangered Resources Branch, Resource Protection Division, Texas Parks and Wildlife Department, Austin (revised 1993)

Markin GP, O'Neil J, Dillier J, Collins HL (1974) Regional variation in the seasonal activity of the imported fire ant, *Solenopsis saevissima richteri*. Environ Entomol 3:446–452

Nature Conservancy (2000) Red imported fire ant: *Solenopsis invicta* Buren. Wildland Invasive Species Program of Nature Conservancy, Arlington

Reddell JR (2001) Cave invertebrate research on Fort Hood, Bell and Coryell Counties, Texas. Report to Texas Nature Conservancy, Fort Hood Field Office, pp 278

Taber SW (2000) RIFA. Texas A&M University Press, College Station

Tana T (2002) Hazard identification and import release assessment: the introduction of red imported fire ants into New Zealand via the importation of goods and arrival of craft from Australia, the Caribbean, South America, and the USA. Biosecurity Authority (now known as Biosecurity NZ), Ministry of Agriculture and Forestry, Wellington

Taylor SJ, Krejcac JK, Smith JE, Block VR, Hutto F (2003a) Investigation of the potential for red imported fire ant (*Solenopsis invicta*) impacts on rare karst invertebrates at Fort Hood, Texas: a field study. Technical Report for Illinois Natural History Survey, Center for Biodiversity, Urbana

Taylor SJ, Sprouse PS, Hutto F (2003b) A survey of red imported fire ant (*Solenopsis invicta*) distribution and abundance at Fort Hood, Texas. Technical Report for Illinois Natural History Survey, Center for Biodiversity, Urbana

Tschinkel WR (1993) Sociometry and sociogenesis of colonies of the fire ant *Solenopsis invicta* during one annual cycle. Ecol Monogr 63(4):425–457

United States Fish and Wildlife Service (2000) Endangered and threatened wildlife and plants: final rule to list nine Bexar County, Texas invertebrate species as endangered. Fed Regist 65(248):81419–81433

United States Fish and Wildlife Service (USFWS) (1994) Recovery plan for endangered karst invertebrates in Travis and Williamson counties, Texas. United States Fish and Wildlife Service, Albuquerque

Wilensky U (1999) NetLogo. Computer software. Center for Connected Learning and Computer-Based Modeling, Northwestern University, Evanston. http://ccl.northwestern.edu/netlogo/. Accessed 05/2009

Wojcik DP, Allen CR, Brenner RJ, Forys EA, Jouvenaz DP, Lutz RS (2001) Red imported RIFA: impact on biodiversity. Am Entomol 47(1):16–23

Chapter 5
Spatially Explicit Agent-Based Model of Striped Newt Metapopulation Dynamics Under Precipitation and Forest Cover Scenarios

Jennifer L. Burton, Ewan Robinson, and Sheng Ye

5.1 Background

The striped newt (*Notophthalmus perstriatus*) is a rare salamander species occurring only in southern Georgia and northern Florida. The newt breeds only in temporary ponds, which can remain dry over many years, within upland sandhill habitats (Dodd 1993). It has a complex life cycle that involves aquatic egg and larval stages, terrestrial juvenile eft and adult forms, and a neotenic aquatic adult (Johnson 2002). Due to land development in the newt's home range, the species is currently restricted to a few locales that contain much of the remaining suitable habitat. Military installations, including Fort Stewart in southeastern Georgia, contain newt populations. It is believed that long-term droughts threaten the newt, which only breeds successfully when sufficient water is present in breeding ponds (Dodd 1993). Climate change and continuing habitat disturbance and fragmentation, including more intense use of bases for military training, further threaten the striped newt. As a result of these factors, the US Fish and Wildlife Service is considering listing the species as endangered. Endangered status would be especially costly to military installations because it would require curtailing training and other land uses around breeding sites. Conservation of striped newt populations is thus a high priority for base staff.

Few detailed studies of striped newt ecology have been performed (Johnson 2002). The co-generic red-spotted newt (*Notophthalmus viridescens*), however, has been studied throughout the eastern United States. The conservation priority of the

J.L. Burton (✉)
Department of Natural Resources and Environmental Sciences, University of Illinois,
W-503 Turner Hall, 1102 South Goodwin Avenue, Urbana, IL 61801, USA
e-mail: jlburton@illinois.edu

E. Robinson • S. Ye
Department of Geography, University of Illinois,
220 Davenport Hall, 607 S. Mathews Avenue, Urbana, IL 61801, USA

J.D. Westervelt and G.L. Cohen (eds.), *Ecologist-Developed Spatially Explicit Dynamic Landscape Models*, Modeling Dynamic Systems,
DOI 10.1007/978-1-4614-1257-1_5, © Springer Science+Business Media, LLC 2012

striped newt means that more information is urgently needed both to inform future studies and to guide management and restoration. This study aims to address this gap by using spatially explicit, agent-based modeling to synthesize existing information on the striped newt and red-spotted newt and to identify priority variables for research and management.

Few systematic ecological studies of the striped newt have been undertaken. The several available studies of the newt's basic life history provide a rudimentary picture of its seasonal movement and habitat use (Dodd 1993, 1997; Dodd and Johnson 2007; Johnson 2002, 2003). Newts reproduce in temporary ponds, where larvae metamorphose into the eft life stage. After approximately 6 months, most efts leave the ponds and move to surrounding upland habitat. When pond conditions are favorable, however, some efts may remain in the pond for another 6 months, metamorphosing into sexually mature paedomorphs and breeding. Paedomorphs then lose their gills and move into the upland (Johnson 2002; Tuberville, February 19, 2009, personal communication). Previous studies have not recorded the ratio of paedomorphs to efts under particular conditions. After migrating, newts remain in the upland until a precipitation event fills the pond once again.

Because temporary ponds are the sole breeding sites for striped newts, pond conditions are likely to have a dominant effect on population viability (Dodd and Johnson 2007). Climatic factors are centrally important, especially the timing and quantity of precipitation, as newts require ponds to contain water for at least 6 months in order to breed successfully (Dodd 1993).

Newt movement includes both micro-movements in and out of ponds and the surrounding area, and migration between ponds and upland habitat. Migration occurs over prolonged periods, with in- and out-migration overlapping. Peaks, during which a maximum number of individuals move, strongly correlate with large rainfall events. One study in northern Florida observed four peaks, the largest occurring between late winter and early summer. The quantity of precipitation was not, however, correlated with the size of the migration (Johnson 2002). Female and male adults normally share similar movement patterns, with both sexes orienting toward several primary directions when leaving and reentering ponds. Juveniles' movement, however, is more multidirectional. Most individuals remain within a small area of upland throughout the dry period (Johnson 2002; Semlitsch 2008). After their first migration away from the pond, efts may disperse to other nearby ponds. Mature newts that have bred previously, however, remain essentially entirely faithful to a single breeding pond (Dodd 1997; Tuberville, February 19, 2009, personal communication).

While there is little literature on the striped newt, much work has been conducted on several closely related subspecies of red-spotted newt (Gabor et al. 2000; Regosin 2005). This species shares a similar life history, migrating from ponds to upland forest habitat in winter. Newt movement is similarly concentrated around breeding ponds, with 83–87% of the newts found within 100 m in one study (Regosin 2005). Adult newts disperse around their natal ponds, usually within 500–600 m (Roe and Grayson 2008) but is known to occur up to 709 m from ponds (Dodd and Johnson 2007). This movement distance is independent of age and sex (Roe and Grayson 2008). As with striped newts, red-spotted newt adults exhibit stronger directionality during migration than do metamorphs (Malmgren 2002); adults are more likely to move through uplands in a direct path than juveniles (Roe and Grayson 2008).

Red-spotted newts prefer forest sites adjacent to breeding ponds (Malmgren 2002). They also prefer upland habitat with significant forest cover. Two studies found no newts in sites with less than 25% (Porej et al. 2004) and 50% forest cover (Gibbs 1998), respectively. Within uplands, red-spotted newts are present beneath leaf litter (40.2%) or branches (29.5%); some also rest on exposed surfaces (16.4%) or ferns and logs (13.9%) (Roe and Grayson 2008).

Red-spotted newts have high dispersal tendency and habitat specificity, and are sensitive to environmental fragmentation and other anthropogenic impacts (Gibbs 1998). Populations are impacted by road construction, controlled burning, and habitat fragmentation (Malmgren 2002). The likelihood of newt population presence in ponds decreases with cumulative road length, distance to nearest wetlands, and increases with amount of forest (Porej et al. 2004). Forest cover may be especially important during droughts, when newts require shade to prevent dehydration (Rohr and Madison 2003). Clearcutting reduces leaf litter mass and depth, driving newts from the area. Canopy fires destroy 50% of the overstory and most of the understory, and also can evaporate ponds and alter water quality (Sadinski and Dunson 1992; Gamradt 1997; Kerby and Kats 1998). While newts may be exposed to herbicides applied near to ponds, recent research has not found a clear impact of herbicide exposure on amphibian population viability or movement (McComb et al. 2008; Cole et al. 1997).

Research on striped newt population dynamics is critical given the species' high conservation priority and the scarcity of existing data. However, the lack of information makes it difficult to choose which variables to assign the highest priority for study. There is thus a need to use the available data to predict which variables may control newt population dynamics in order to guide future research. For this type of situation, spatially explicit agent-based modeling is a method well suited for informing research because sensitivity analysis can be used to identify the most important variables even where data may be unavailable. At the most basic level, an agent-based model (ABM) can assess the relative importance of mortality in upland vs. pond habitats for overall population viability. Thus, sensitivity analysis can help managers decide whether to prioritize upland habitats or pond habitats in their future research on competition, predation, and other environmental factors affecting mortality. While available research demonstrates that striped newts exhibit different behavior and movement patterns at different life stages, little is known about how mortality rates change across these stages. An ABM can assess whether mortality in one life stage is especially important to metapopulation viability.

Because of the importance of temporary breeding ponds to striped newt populations and the predicted decline in rainfall in the southern United States with climate change (Burkett et al. 2001), the effects of drought are a central component of population viability (Dodd 1993; Johnson 2002). Agent-based modeling allows us to examine the viability of striped newt populations in several pond clusters under different climatic scenarios.

Upland habitat mosaics are another important determinant of newt populations documented in the literature, and one that can be altered by human intervention.

Agent-based modeling allows the assessment of different upland habitat scenarios in conjunction with climate scenarios in order to quantify the impact of potential habitat management through, for example, controlled burns.

5.2 Objective

In order to identify priorities for future field studies, we developed a spatially explicit ABM to identify the most important variables impacting the dynamics of striped newt metapopulations in the southeast United States. When resources are limited, factors deemed to have the highest impact on simulated newt population viability may be the best candidates for field study. Then, targeted field research based on model results may quantify key dimensions of newt ecology and inform management approaches aimed to ensure long-term metapopulation viability.

We hypothesize that precipitation will be the controlling factor for newt metapopulation dynamics. Specifically, we hypothesize that reduced monthly precipitation will have the greatest negative impact on newt populations among the variables tested. Increased seasonal and interannual variability will also substantially increase the risk of newt extirpation. Furthermore, we hypothesize that the newt population will be sensitive to the percentage of forest canopy cover across the landscape. However, we hypothesize that forest fragmentation will have a less pronounced impact than precipitation.

5.3 Model Description

5.3.1 Purpose

The model allows users to quantify and compare the relative impacts of broad-scale variables on newt metapopulations. Changes in input values for climatic and habitat variables are expected to alter the long-term viability of the simulated newt metapopulation. Quantification of these changes makes it possible to assess the relative importance of certain variables in areas where published data are scarce.

5.3.2 Modeling Tools

We used NetLogo 4.04 (Wilensky 1999) to develop the model.[1] ArcGIS spatial data were imported into NetLogo, which resampled rasters into 100×100 m patches.

---

[1] An operational copy of this model is available through http://extras.springer.com.

5.3.3 Data Sources

We used public domain climate data and the following vector map layers for the Fort Stewart area: elevation contours, known striped newt ponds, canopy cover, and roads. The elevation contours were converted to a raster-based digital elevation model (DEM); the drainage networks, pond area, and catchment contribution area were extracted using ArcGIS.

5.3.4 Study Area

Fort Stewart, adjacent to Savannah, GA in the southeastern United States, provides the geospatial context for this study. This model was developed for a 640 ha area in the forested northwestern portion of the installation. Striped newt ponds tend to be located at the highest areas of watersheds in conjunction with Ellabelle loamy sand areas. This suggests that newts prefer extremely ephemeral ponds. A rainfall event sufficiently heavy to fill these ponds may occur only once in several years. These areas are depressions associated with very small watersheds. The extremely ephemeral nature of these pond areas may result in very low aquatic predation and competition. Adapting to survive spans of several years without reproductive events may result in a minimization of predators (Semlitsch 2008).

5.3.5 State Variables and Scales

At the most basic level, the model is composed of agents, patches, and state variables. Agents represent individual striped newts. Two types of newt agents exist in the model: efts (terrestrial juveniles) and adults. After 12 months, efts metamorphose into adults. Patches represent two-dimensional 100×100 m areas of the landscape. The spatial extent of the landscape is an area of $1,500 \times 1,700$ m. Agents that move beyond this area are removed from the model simulation. Patches are divided into two categories according to the behaviors that the newts perform in them: (1) depressions, which can serve as breeding habitat when they contain water; and (2) upland habitat, through which newts move, and in which they settle when not breeding (Fig. 5.1). State variables (properties possessed by agents and patches) and the units to which they apply are listed in Table 5.1. They are described in greater detail below.

Lifestage dictates whether a newt agent is an adult or an eft. These two agent types behave differently. Again, efts metamorphose into adults after 12 months.

Behavior trigger determines which actions are currently being undertaken by a newt. The variable carries four values: "move into upland," "move to pond," "upland," and "breed."

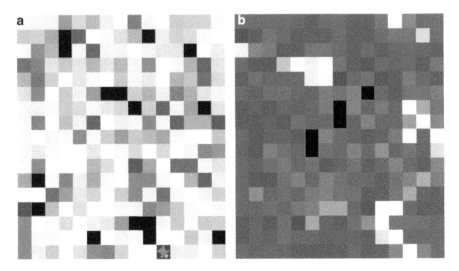

Fig. 5.1 Model landscapes as displayed in the NetLogo environment. *Shading* represents the percentage of canopy cover in a patch, with darker patches having more cover. *Black patches* represent depressions that can hold ephemeral breeding ponds. (**a**) Simulated landscape. (**b**) Landscape based on raster data for an area of Fort Stewart, GA

Table 5.1 State variables used in the model, according to the unit with which they are associated (agents or patches)

Newt properties		Patch properties	
Variable	Type	Variable	Type
Life stage	Categorical	Patch type	Categorical
Behavior trigger	Categorical	Canopy cover	Percentage
Home pond location	Integer set	Depth	Integer
Rainfall sensitivity	Integer	Catchment area	Integer
		Pond area	Integer

Home pond location stores the Cartesian coordinates defining the pond in which a newt breeds. Efts do not have a home pond location until they mature into adults, after which point they select the closest pond as their home pond. Adults are 100% loyal to a single home pond throughout their lifetime.

Rainfall sensitivity defines the minimum quantity of monthly rainfall that will trigger a newt to return to its home pond and attempt to breed. Only newts whose current behavior trigger is set to "upland" respond to precipitation in this way. The values of rainfall sensitivities are normally distributed across newts, with the model user determining the mean value and standard deviation. The rainfall sensitivity of a particular newt agent does not change throughout its lifetime.

Patch type dictates whether a patch is a depression capable of forming an ephemeral breeding pool or an upland forest.

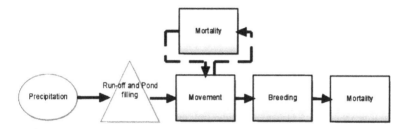

Fig. 5.2 Sequence of model processes. Precipitation is read from user-determined input data (*oval*); runoff and pond filling are landscape processes (*triangle*); the remaining processes are agent-based activities (*rectangles*). All processes occur in discrete time steps, except movement and movement-driven mortality, which are recursive (indicated by *dashed lines*)

Canopy cover defines the percentage of canopy cover for a given patch. Depressions are defined as having a 0% canopy cover.

Depth represents the current water depth in a depression patch in meters. Depressions are assumed to be "bucket-shaped," with a uniform depth across their area.

Catchment area describes the surface area draining into a depression. This property only applies to depressions, not to upland patches. Note that if the user chooses to use Fort Stewart geographic data, then catchment area is measured in square meters, but if the user selects a simulated landscape, then catchment area is measured in numbers of patches.

Pond area is the surface area of depression patches. When the Fort Stewart landscape is used, this value is defined by measurements made using ArcGIS. In the simulated landscape, depressions are assumed to exactly fill 100×100 m patches.

Each time step in the model is equivalent to 1 month, in order to allow newts to enter and exit the ponds in overlapping waves. The time horizon is 287 months, corresponding to the period over which observed precipitation data are available for Fort Stewart. A 1-month time step does not account for newt micro-movements within the pond, the upland, or between the pond and bank habitat. However, we judged such micro-movements unimportant to overall metapopulation viability.

5.3.6 Process Overview and Scheduling

The processes making up the model and the order in which they occur are depicted in Fig. 5.2. All processes except for newt movement and mortality occur in discrete 1-month time steps. The newt movement function is recursive: each newt moves into one patch, determines whether to remain in the patch it enters, and, if it decides to continue, the newt performs the movement function again. Because newts are subject to a risk of mortality each time they enter a patch, the mortality process is also recursive.

5.3.7   Design Concepts

5.3.7.1   Emergent Phenomena

Spatial distribution of newts across the landscape is the most apparent emergent property within the model. The distribution of newts around ponds is the result of individual newt choices based on the canopy cover of surrounding patches and the distribution of rainfall sensitivities across the newt population. The distance which newts (especially efts) disperse away from ponds is a critical factor in determining rates of recolonization of ponds whose newt population has been locally extirpated (Petranka and Holbrook 2006; Semlitsch 2008). Dispersal distance may thus have important implications for striped newt conservation. Furthermore, the distribution of newt distances from home ponds can serve as a means of evaluating the model, since previous studies provide data on newt dispersal distance (Regosin 2005; Dodd and Johnson 2007) and directionality (Dodd and Cade 1998; Roe and Grayson 2008). Further research rigorously quantifying newt distribution across upland habitat may help to validate and calibrate the model.

5.3.7.2   Adaptation

Intergenerational selection is not present in the model. Adaptation alters the distribution of sensitivities to rainfall across the newt population only insomuch as newts that respond to very low precipitation amounts are more likely to return to ponds containing water insufficient for successful breeding. These individuals then suffer additional risks of mortality when they return to the upland. However, in the model there is no relationship between the rainfall sensitivity of adults and their offspring.

5.3.7.3   Stochasticity

The NetLogo random number generator is used in several model processes, including movement, mortality, and breeding. Stochasticity is also present in the distribution of rainfall sensitivity across the newt population, which is based on a random normal function with the user determining the mean and standard deviation values. When a simulated landscape is used, a random canopy cover value is assigned to each patch. The canopy value is modified by the user-selected canopy modification index (see Sect. 5.3.8.2 below).

5.3.7.4   Observation

The user interface includes plots of the total number of newts, subdivided among adults and efts. Other plots depict the mean depth of ponds, mean and maximum newt distance to home pond, histogram of newt distances to home ponds, and variable mortality rate (when precipitation-influenced mortality is used). All plots are calculated based on values at the end of each time step. For the purpose of analysis, the maximum, minimum, and mean newt populations were used. However, in the initial round of analysis, only the total number of newts at the end of the time horizon was used.

5.3.8   Initialization

Because virtually no quantitative data were available on striped newt ecology, we used two approaches to generate the initial model state: randomization and user-specification of variables. Two dimensions of the initial model state are based on randomization: the geographic distribution of newts across the landscape and the distribution of canopy cover across patches. The function for determining newt positions at the start of the model is described below under Sect. 5.3.8.1. Canopy cover is only randomly generated if the user chooses to employ a simulated landscape; otherwise, cover values from the Fort Stewart study site are used. Furthermore, the user can alter the distribution of canopy cover across the simulated landscape using the canopy modifier slider. The landscape simulation process therefore combines randomization and user specification. For the sake of clarity, it is therefore described in more detail below under Sect. 5.3.8.2.

5.3.8.1   Randomized Initialization

At the beginning of the model, newts are distributed in upland patches surrounding ponds such that each newt has an equal probability of "landing" in patches less than or greater than three patches from a depression. Furthermore, newts are more likely to land in patches with greater canopy cover, and they are distributed independently of one another. The two mathematical criteria used are as follows:

1. $cover^2$ > random integer falling between 0 and 10,000
2. $Distance\ to\ depression = 3 \pm$ random integer falling between 0 and $\sqrt{10}$

   It should be noted that the canopy cover criterion is identical to that underlying newt movement as discussed under Sect. 5.3.10.

5.3.8.2   User-Specified Initialization

Before running the model, the user specifies values pertaining to three model components: the landscape, agent behavior, and landscape–agent interactions (Table 5.2).

Table 5.2 User-determined variables

Landscape variables	Agent variables	Agent–landscape interaction variables
Number of ponds	Initial number of newts	Upland carrying capacity
Canopy cover multiplier	Maximum recruitment per adult	Pond carrying capacity
Leakage rate	Mean and standard deviation of rainfall sensitivity	
Evapotranspiration time		

The user selects one of two methods for determining landscape initial state: reading in observed spatial data from Fort Stewart, as described previously, or generating a simulated landscape. In the simulation option, the user specifies three variables that determine the makeup of the landscape: the number of pond depressions, the mean catchment area of ponds (in number of patches), and the canopy cover multiplier. Canopy cover in the simulated landscape is generated by a random cover value (between 0 and 100) that is assigned to each patch. These cover values are then modified according to the user-specified canopy cover multiplier. If the user selects a negative value, the cover of each patch is decreased by a random integer falling between 0 and the forest cover multiplier. If the user selects a positive value, the cover of each patch is increased in this same manner. No patch can have a canopy value greater than 100 or less than 0.

The user also specifies several terms underlying landscape hydrology, as only limited hydrological data were available for the study site. The model employs a user-determined parameter to alter the linear function governing leakage from ponds. When this value is increased, pond depth has a greater impact on the rate of water leakage from depressions. The user also selects one of two ways for calculating evapotranspiration: either a linear function similar to the leakage function, or a calculation of evapotranspiration based on Dalton's law of partial pressures (Meyer 1942). The hydrological functions used in the model are described in greater detail under Sect. 5.3.10.

The agent factors specified by the user include the number of individuals at the start of the model, the maximum number of efts recruited per breeding adult, and the mean and standard deviation of newt rainfall sensitivity (see previous discussion under Sect. 5.3.5). The user also chooses the carrying capacities of upland and pond patches.

5.3.9 Input

The user selects one of six precipitation scenarios as inputs into the model. The first scenario (a) comprises observed monthly precipitation for Fort Stewart between 1985 and 2008 (Fig. 5.3). The second (b) and third (c) scenarios represent reduced rainfall potentially associated with climate change; they correspond to monthly

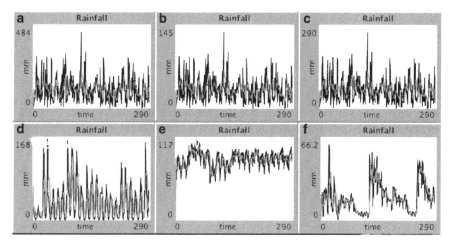

Fig. 5.3 Precipitation data used in the model for 287 months. (**a**) Observed precipitation (1985–2008), (**b**) 40% of observed precipitation, (**c**) 60% of observed precipitation, (**d**) high seasonal fluctuations, (**e**) low seasonal fluctuations, and (**f**) high seasonal and interannual fluctuations

precipitation of 40 and 60% of observed rainfall values, respectively. The final three scenarios represent altered precipitation seasonality with the same total precipitation falling during the study time horizon. In the first of these (d), intra-annual variability is increased, with more rain falling during the winter months. In the second (e), intra- and interannual variability is reduced, with relatively similar amounts of precipitation falling in all months. In the final scenario (f), intra- and interannual variability is increased, producing wet and dry months, as well as several-year periods with little rainfall.

5.3.10   Submodels

5.3.10.1   Pond Hydrology

The model treats pond depressions as "buckets": water enters through subsurface and surface flow from catchment areas and leaves through evapotranspiration and leakage. Incoming flow is the result of rainfall. The quantity of water stored in a pond in a given month is calculated as follows:

Pond water volume in previous month + precipitation – evapotranspiration – leakage

Precipitation is read from input files, and leakage is assumed to be a linear function of water storage in pond—the more water in the pond, the more quickly leakage occurs. The rate factor for this linear function can be changed based on the users' understanding of soil properties using the variable "leak-time."

The user can specify one of two mechanisms for determining evapotranspiration. The first is the same as the leakage mechanism: evapotranspiration is calculated as a linear function of pond water storage. As with leakage, the user can specify the residence time based on his or her understanding of local climate, using the variable "evap-time." The second evapotranspiration mechanism uses the empirical formula developed by Meyer (1942) based on Dalton's law, which is suitable for calculating evapotranspiration for open water bodies. The formula is as follows:

$$E = C(e_s - e_d)\left(1 + \frac{u25}{10}\right)$$

where E is evapotranspiration; C is the coefficient, equal to 15 for shallow ponds; e_s is the saturation vapor pressure of air at the water temperature at 1 ft deep; e_d is the actual vapor pressure of air equal to e_s*relative humidity; and $u25$ is the average wind velocity at 25 ft above the pond surface. e_s, e_d, and $u25$ are adapted from the average value posted online. These variables can be changed in the model code, but are not present in the user interface.

5.3.10.2 Movement

As previously discussed, newt movement, unlike the other submodels, is a recursive process occurring continuously within discrete 1-month time steps. There are also two distinct types of movement: (1) migration away from ponds and into the upland following breeding, and (2) returning to home ponds after rainfall of sufficient magnitude.

When moving away from ponds, newts move in single-patch increments, and determine after each increment whether to settle in the current patch or move again. The process occurs as follows:

1. Newts rotate to face a random heading, then move into the patch directly in front of them.
2. After entering the patch, a risk of mortality is applied based on the canopy cover of the new patch (see Sect. 5.3.10.3 below).
3. If they survive, they then decide whether to remain in the patch, again based on the patch's canopy cover.

The criterion for which newts will settle in the patch is:

$cover^2$ > random integer falling between 0 and 10,000

The second power of canopy cover is used in order to produce increasing marginal likelihoods of newt settling as patch cover increases. This relationship was selected because it produced model behavior best matching the observations that striped newt and red-spotted newt individuals do not occur in areas with less than 50% canopy cover (Gibbs 1998).

When newts return to ponds from the upland, a second and distinct movement process operates. Newts jump from the upland patch in which they settled directly into their home pond patch. The move occurs instantaneously and disregards the canopy cover values (and all other properties) of patches that the newt crosses.

5.3.10.3 Mortality

Mortality occurs in three ways during each time step. First, newts moving away from ponds and into the upland face a risk of mortality for each patch of upland habitat they cross. Second, newts leaving the upland and returning to ponds are subject to a one-time risk of mortality. This introduces a tradeoff for newts with low rain sensitivity; they are more likely to return to ponds to breed following rainfall, but are also subject to higher risks of mortality. The third type of mortality affects all newts in upland patches. This simulates attrition while newts are "settled" in the upland, increasing the risk of mortality for newts that settle in patches with lower canopy cover. The criterion for determining the risk of mortality at all three of these stages is the same:

$$\text{Random integer between 1 and } 10 > \frac{\sqrt{\text{cover}}\left(1 - \dfrac{n_t}{N_{max}}\right)}{\text{mortality} \sqrt[4]{\dfrac{\hat{p}}{\overline{p}}}}$$

where cover is the percentage canopy cover of the upland patch, n_t is the number of newts in the patch at the current time step, N_{max} is the user-determined carrying capacity for upland patches, mortality is the user-determined mortality coefficient, $\hat{p}$ is the user-determined base precipitation rate, and $\overline{p}$ is the 4-month mean precipitation.

Since very little data were available on the determinants of striped newt mortality, the above approach requires the user to select mortality criteria based on his or her understanding of newt ecology. The numerator of the criterion represents the ecological effects of habitat and intraspecific competition. The square root of cover is used in order to reduce the marginal effect of a 1% increase in canopy cover. This allows a substantial percentage of newts occurring in patches with greater than 50% canopy cover to persist for at least several time steps. The second term in the numerator represents standard density-dependent mortality. The denominator comprises two components that serve to alter mortality risks. First, the mortality coefficient allows the user to alter mortality rates uniformly across all newts for the entire simulation. Second, precipitation-based mortality can be turned on or off depending on the extent to which the user believes rainfall influences newt survival in uplands. Turning on precipitation-based mortality causes the mortality rate to fluctuate with the 4-month mean of precipitation. When the mean exceeds the user-selected base precipitation level, the mortality rate is decreased for all newts; when the mean is less than this level, mortality is increased.

It should be noted that newts are not subject to the mortality process while they are in ponds. We excluded in-pond mortality due to the paucity of available data and because we believed breeding was the controlling factor in newt populations while they are in the pond (see Sect. 5.3.10.4 below).

5.3.10.4 Breeding

Breeding occurs when newts return to their "home ponds" following monthly rainfall events exceeding their individual rainfall sensitivities. Because very little data were available on newt reproduction, the model employs a highly simplified reproduction process. The model excludes the larvae life stage. Instead, reproduction produces newts in the eft life stage that are ready to leave the pond and move into the upland. Water must be present in the pond depression for at least 6 months in order for larvae to survive and metamorphose into efts (Johnson 2002). Thus, in order to breed successfully in the model, newts must be present in a pond for 6 months during which water depth does not fall to zero. If pond depth is zero during any month in which newts are in the pond, breeding fails and the newts move into the upland using the movement process (Sect. 5.3.10.2) described above. If, however, pond depth remains above zero, each newt produces a number of efts determined by the function:

$$\text{Number of efts recruited} = \overline{\text{depth}} * \text{recruit}_{max} * \left(1 - \frac{n_t}{N_{max}}\right)$$

where $\overline{\text{depth}}$ is the mean pond depth during the 6-month breeding period, recruit_{max} is a random integer less than the user-defined maximum recruitment per breeding adult, and the final term represents standard density dependence.

5.4 Simulation Experiments

5.4.1 Model Calibration

Long-term population dynamics of the striped newt are unavailable. Therefore, in lieu of statistically based validation, we adopted a qualitative approach (Rykiel 1996). We repeatedly compared model behavior under different conditions to a priori expectations. For example, literature on pool-breeding amphibians indicates that while population levels peak after breeding events, effective breeding populations are very much smaller (Johnson 2002). We therefore adjusted breeding and mortality functions in order to reduce observed instances of exponential population

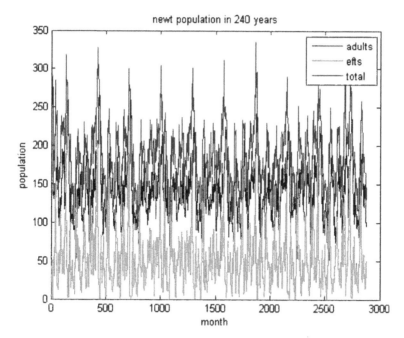

Fig. 5.4 Newt populations in a single simulation conducted over 2,870 months generated by looping the 287-month time horizon of observed precipitation data

growth and rapid extinction across the study period. In addition, we calibrated the model with regard to observed patterns of pond filling and drying and newt migration to and from breeding ponds in response to rainfall (Johnson 2002).

Under conditions of limited data availability, we followed the principle of creating the minimum amount of complexity necessary to generate qualitative expectations. Where more than one function appeared to produce realistic behavior, the least complex was used in the model.

The normal values for user-determined variables (the mid-point on the model interface slider bars) were similarly determined to guard against the qualitative extremes of system behavior. The maximum and minimum values were selected such that they tend to produce behavior impossible given real world conditions. Therefore, they represent the limits of the range most likely to interest model users for each parameter.

One concern that arose during calibration was whether the time horizon over which precipitation data were available was sufficiently long enough to produce equilibrium behavior. Therefore, we produced long-term time horizon precipitation data by looping observed data ten times, producing a 2,870 month dataset. Visual inspection of population plots for this dataset shows that population fluctuations do not appear to change over long-term periods of time (Fig. 5.4). This suggests that the system is in steady state during the first loop of precipitation data.

Table 5.3 Coefficients of correlation between end population in tests and under control conditions

Variable	Coefficient of correlation at value	
	50%	150%
Landscape		
Precipitation[a]	0.54	0.69
Evapotranspiration time	0.41	0.78
Leakage time	0.54	0.66
Canopy cover	0.59	0.83
Number of ponds	0.71	0.88
Catchment size	0.72	0.8
Agent		
Maximum recruitment/adult	0.73	0.89
Rainfall sensitivity	0.73	0.56
Rainfall sensitivity standard deviation	0.91	0.9
Ecology		
Pond carrying capacity	0.57	0.82
Mortality coefficient	0.58	0.66
Mortality precipitation rate	0.95	0.93

Tests scenarios involved setting each variable individually to 50 or 150% of its "normal" value, holding other variables constant. Low coefficients of correlation indicate that the model is more sensitive to a variable at the given value
[a]The high precipitation value is 167% of the "normal" value, and thus its correlation coefficient cannot be compared to others

5.4.2 Preliminary Results

We tested the sensitivity of the model to changes in the values of each variable individually. Control conditions constituted all variables being set to their mid-point values, which were selected during calibration. Then, each test constituted a single variable being set to 150 or 50% of its mid-point value, holding all other variables constant. All tests were conducted using 60 trials of the 287-month time horizon. The total newt population at the end of each trial was used as the output. We did not use measures of population variability or newt spatial distribution in preliminary analysis. Correlation coefficients between final newt population under test conditions and final population under control conditions were used as measures of sensitivity. Results were interpreted in the following manner: lower coefficients of correlation indicate that the model is more sensitive to the variable at its given value (50 or 150%).

Results indicate a range of sensitivities, with correlation coefficients varying from 0.56 to 0.93 using the 150% values, and from 0.41 to 0.95 using the 50% values (Table 5.3). Among the 150% conditions, newt population was most sensitive to mean "sensitivity to rainfall." Among the 50% conditions, newt population was most sensitive to precipitation scenario (Fig. 5.5).

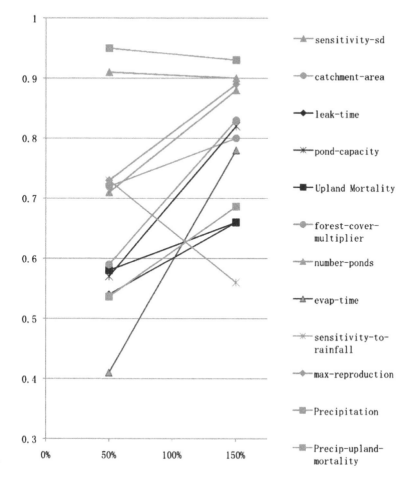

Fig. 5.5 Coefficients of correlation between end populations in the control scenario and test scenarios. Each variable was tested individually

Because rainfall was hypothesized to be a central factor in controlling newt populations, we tested model sensitivity to the five precipitation scenarios. The same approach was used: based on 60 trials, we calculated coefficients of correlation between the constructed scenarios and observed precipitation data, using newt population at the end of the study period as the output variable (Table 5.4).

Two caveats should be noted in interpreting the above data. First, we used observed precipitation data from Fort Stewart as the high precipitation value. The 60% scenario was used as the normal value, and the 30% scenario was used as the low value. The high value is thus 167% of the normal value, rather than 150%. Therefore, the correlation coefficient at the high value is not directly comparable with those of other variables. Second, there are three variables that represent properties of the simulated landscape but do not apply to the Fort Stewart study

Table 5.4 Coefficients of correlation between end population in precipitation scenarios and in observed data

Scenario	Coefficient of correlation
30% Precipitation	0.37
High seasonal variation	0.42
Low variation	−0.45
High interannual variation	0.5
60% Precipitation	0.69

Low values indicate the model is substantially sensitive to the scenario

area: canopy cover, number of ponds, and catchment size. The correlation coefficients for these variables should therefore not be directly compared with others.
These limitations notwithstanding, comparison of correlation coefficients gives an
initial sense of the importance of various ecological processes in determining newt
population dynamics and viability.

5.5 Discussion

Preliminary sensitivity analysis suggests that several processes may strongly determine newt population dynamics. At both high and low values, three variables—
leakage rate, upland mortality coefficient, and precipitation—showed correlation
coefficients of less than 0.70. Furthermore, two variables had little impact at both
high and low values, with correlation coefficients equal to or greater than 0.90:
variation in sensitivity to rainfall and precipitation-based mortality. These results
suggest three areas for further field study of striped newts:

1. Factors affecting leakage rates from different depressions, which are likely linked
   to soil type.
2. Factors influencing mortality as newts both move through and settle in upland
   forests. Precipitation-related mortality may be a less important factor than other
   sources of mortality, as the model was least sensitive to the base precipitation
   rate at both high and low values.
3. Predicted precipitation trends associated with climate change in the study area.
   Changes in precipitation, especially should they drastically reduce monthly rainfall or increase seasonal and interannual variation, may strongly impact newt
   population dynamics (Table 5.2).

Priorities one and three relate to a priori expectations that pond hydroperiods
critically control striped newt populations (Dodd 1993; Tuberville, February 19,
2009, personal communication). Priority two suggests that upland sources of mortality cannot be ignored.

The differences between variable effects at low and high values could provide a
basis for further model development or field research. Three variables displayed

Table 5.5 Changes in correlation coefficients between high and low value tests of sensitivity

Rank	Variable	Change
1	Evap-time	0.37
2	Pond-capacity	0.25
3	Forest-cover-multiplier	0.24
4	Number-ponds	0.17
5	Sensitivity-to-rainfall	−0.17
6	Max-reproduction	0.16
7	Precipitation	0.15
8	Leak-time	0.12
9	Upland mortality	0.08
10	Catchment-area	0.08
11	Precip-upland-mortality	−0.02
12	Sensitivity-sd	−0.01

coefficients that differed by more than 0.20 across low and high values: evapotranspiration rate, pond carrying capacity, and forest cover (Table 5.5). All of these variables had a correlation coefficient of less than 0.70 at either their high or low value, suggesting that they may, under some conditions, act as important determinants of newt population dynamics.

5.6 Conclusions

The simulation results support our hypothesis that, based on available information, rainfall is likely a key factor governing striped newt population viability. The model highlighted the importance of hydrologic variables such as precipitation, leakage from ponds, and evapotranspiration, and it suggested that pond depth may be a key factor controlling newt populations. Regarding our second hypothesis, the results do not provide conclusive evidence regarding the importance of forest canopy cover; while decreasing cover had a substantial negative effect on newt populations, increasing cover had little effect.

While tests of model sensitivity to joint and contingent conditions are needed in order to fully specify model implications for research on the striped newt, priorities for future fieldwork might include characterizing rates of leakage and evapotranspiration in pond depressions and describing the relationship between pond depth, breeding, and eft recruitment. More detailed analysis of climate change scenarios may also be warranted; the precipitation scenarios used in this report represent only a few possible scenarios (and simplified ones) for changes in rainfall quantity and variability. The current model design allows for further precipitation data to be incorporated in a straightforward manner.

Our analysis has treated comparisons in a qualitative and preliminary way. Further analysis should measure the effects of varying multiple variables jointly, and determine the statistical significance of different outcomes.

References

Burkett VR, Ritschard R, McNulty S, O'Brien JJ, Abt R, Jones J, Hatch U, Murray B, Jagtap S, Cruise J (2001) Potential consequences of climate variability and change for the southeastern United States. In: National Assessment Synthesis Team for the U.S. Global Change Research Program (ed) Climate change impacts in the United States: potential consequences of climate variability and change. Cambridge University Press, Cambridge, pp 137–164

Cole EC, McComb WC, Newton W, Chambers CL, Leeming JP (1997) Response of amphibians to clear cutting, burning, and glyphosate application in the Oregon coast range. J Wildl Manage 61(3):656–665

Dodd CK (1993) Cost of living in an unpredictable environment: the ecology of striped newts *Notophthalmus perstriatus* during a prolonged drought. Copeia 3:605–614

Dodd CK (1997) Movement patterns and the conservation of amphibians breeding in small, temporary wetlands. Conserv Biol 12(2):331–339

Dodd CK, Cade BS (1998) Movement patterns and the conservation of amphibians breeding in small, temporary wetlands. Conserv Biol 12(2):331–339

Dodd CK, Johnson SA (2007) Breeding ponds colonized by striped newts after 10 or more years. Herpetol Rev 38(2):150–152

Gabor CR, Krenz JD, Jaeger RG (2000) Female choice, male interference, and sperm precedence in the red-spotted newt. Behav Ecol 11(1):115–124

Gamradt SC (1997) Impact of chaparral wildfire-induced sedimentation on oviposition of stream-breeding California newts (*Taricha torosa*). Oecologia 110(4):546–549

Gibbs JP (1998) Distribution of woodland amphibians along a forest fragmentation gradient. Landsc Ecol 13:263–268

Johnson SA (2002) Life history of the striped newt at a north-central Florida breeding pond. Southeast Nat 1:381–402

Johnson SA (2003) Orientation and migration distances of a pond-breeding salamander (*Notophthalmus perstriatus, Salamandridae*). Alytes 21(1, 2):3–22

Kerby JL, Kats LB (1998) Modified interactions between salamander life stages caused by wild-fire-induced sedimentation. Ecology 79(2):740–745

Malmgren JC (2002) How does a newt find its way from a pond? Migration patterns after breeding and metamorphosis in great crested newts (*Trititus cristatus*) and smooth newts (*T. vulgaris*). Herpetol J 12:29–35

McComb BC, Curtis L, Chambers CL, Newton M, Bentson K (2008) Acute toxic hazard evaluations of glyphosate herbicide on terrestrial vertebrates of the Oregon coast range. Environ Sci Pollut Res 15(3):266–272

Meyer AF (1942) Evaporation from lakes and reservoirs. Minnesota Resources Commission, St. Paul

Petranka JW, Holbrook CT (2006) Wetland restoration for amphibians: should local sites be designed to support metapopulations or patchy populations? Restor Ecol 14(3):404–411

Porej D, Micacchion M, Hetherington TE (2004) Core terrestrial habitat for conservation of local populations of salamanders and wood frogs in agricultural landscapes. Biol Conserv 120:399–409

Regosin JV (2005) Variation in terrestrial habitat use by four pool-breeding amphibian species. J Wildl Manage 69(4):1481–1493

Roe AW, Grayson KL (2008) Terrestrial movements and habitat use of juvenile and emigrating adult eastern red-spotted newts, *Notophthalmus viridescens*. J Herpetol 42(1):22–30

Rohr JR, Madison DM (2003) Dryness increases predation risk in efts: support for an amphibian decline hypothesis. Oecologia 135:657–664

Rykiel EJ Jr (1996) Testing ecological models: the meaning of validation. Ecol Model 90(3):229–244

Sadinski WJ, Dunson WA (1992) A multilevel study of effects of low pH on amphibians of temporary ponds. J Herpetol 26(4):413–422

Semlitsch RD (2008) Differentiating migration and dispersal processes for pond-breeding amphibians. J Wildl Manage 72(1):260–267

Wilensky U (1999) NetLogo: computer software. Center for Connected Learning and Computer-Based Modeling, Northwestern University, Evanston. http://ccl.northwestern.edu/netlogo/. Accessed 01/2011

Chapter 6
Forecasting Gopher Tortoise (*Gopherus polyphemus*) Distribution and Long-Term Viability at Fort Benning, Georgia

James D. Westervelt and Bruce MacAllister

6.1 Background

The goal of this study is to provide information to military installation land managers who make decisions on land-use allocations that affect the viability of species at risk (SAR) populations over decades and centuries. As a department of the US government, the US Army is required both by federal law and Army regulation to ensure the long-term persistence of species residing on its training lands. However, the Army must also ensure that sufficient training and testing areas remain available to meet current and future Army needs. If a SAR residing on Army installations is elevated to federally threatened status, it could compromise military readiness through the associated loss of training land (Guertin 2005). This challenge must be met with careful and cost-effective regional planning that will proactively ensure adequate habitat for SAR in the face of cumulative natural and anthropogenic changes.

The gopher tortoise (*Gopherus polyphemus*) is a charismatic SAR that has historically occupied large areas of the southeastern USA. This tortoise can easily live 50 or more years (Iverson 1980; Landers et al. 1980; Rostal and Douglas 2002) and has a delayed sexual maturity ranging from 12 to 20 years (Mushinsky and McCoy 1994), resulting in a generation time of around 30 years. It is federally protected in the western portion of its range, and is a species of interest throughout its entire range (Fig. 6.1). Federal lands, like Fort Benning, Georgia, are aggressively engaged in maintaining remaining populations at the same time the land is used more completely and intensively. Unfortunately, the preponderance of information about the gopher tortoise comes from a relatively small body of field studies. These studies are informative and significant, but taken together they are analogous to the disjunct surviving frames of a lost feature film: a full understanding of the gopher

J.D. Westervelt • B. MacAllister (✉)
Construction Engineering Research Laboratory, US Army Engineer Research
and Development Center, Champaign, IL, USA
e-mail: james.d.westervelt@usace.army.mil; Bruce.A.MacAllister@usace.army.mil

J.D. Westervelt and G.L. Cohen (eds.), *Ecologist-Developed Spatially Explicit Dynamic Landscape Models*, Modeling Dynamic Systems,
DOI 10.1007/978-1-4614-1257-1_6, © Springer Science+Business Media, LLC 2012

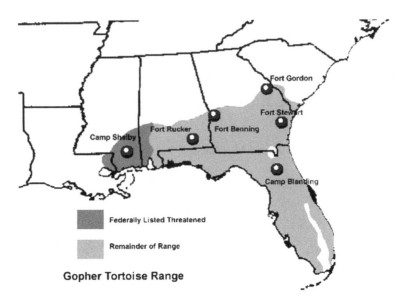

Fig. 6.1 Gopher tortoise range

tortoise based on available field studies is as elusive as understanding the plot of a motion picture based on a collection of film scraps.

A question generally asked when new development is proposed is how is that development likely to change the probability that the local population remains viable throughout the foreseeable future. Answering this question generally requires intimate knowledge of the dynamics of the local populations—especially as part of a larger metapopulation within which individuals can move among the separate populations. A population viability analysis (PVA) involves formally modeling a system to identify the probability that a population will persist through projected human activities and random natural events involving weather, fire, and changes in community structures. A PVA can be developed to forecast the impact of proposed and actual conservation and land management decisions made today. Population viability analyses are often based on knowledge of the behavior of populations using models such as Vortex and RAMAS (Lindenmayer et al. 1995; Akçakaya 2002). Unfortunately, in the case of the long-lived gopher tortoise little is known about the dynamics of populations at this scale. However, much has been studied and is published about the behavior of individuals over months to years. We are therefore limited to using published results of such studies as the foundation for proposing and testing alternative regional planning options designed to strike the balance between the survival of target populations and the needs for human activity. Specific management strategies that require testing include the allocation of land use over space and time and the scheduled management and maintenance of land. Therefore, we look at the potential for using this information to construct a dynamic spatially explicit model that might help evaluate the consequences of proposed land use changes with the anticipation that the behavior of a metapopulation will emerge from published knowledge of individual behavior.

6.2 Objective

The objective of this agent-based spatially explicit simulation model is to test alternative land use allocation and land management strategies with respect to the probability of gopher tortoise population survival, as based on published literature and expert advice.

6.3 Model Description

6.3.1 Purpose

Our research questions involve the allocation and management of land over time, which requires an environment that supports the simulation of processes on a landscape scale and incorporates individual tortoise behaviors over space and time. Modeling is best accomplished when the user of the system is deeply familiar with the model itself. Therefore, we required a modeling environment that allows for constructing models that are easily understood by ecologists. We selected NetLogo (Wilensky 1999), a user-friendly simulation modeling environment based on a computer language of the same name.[1] The NetLogo language is a rich, high-level programming language that is easy to learn and use. NetLogo technology is designed to enable scientists with no computer programming experience to read, write, modify, and distribute highly explanatory simulation models.

This model is intended to be used by land managers to test alternative land management decisions and their long-term consequences with respect to local gopher tortoise populations. In this chapter, we hope to show that there is sufficient published information about the gopher tortoise to develop a spatially explicit individual-based simulation model that could help guide the development of land management plans.

The model was developed for a small area known to support gopher tortoises within Fort Benning, which is located on the border of Georgia and Alabama (Fig. 6.2a and 6.2b). The study site is centered near 32°22′N and 84°42′W and is approximately 3.36×2.76 km in size (Fig. 6.3). The area supports a population of

---

[1] An operational copy of this model is available through http://extras.springer.com.

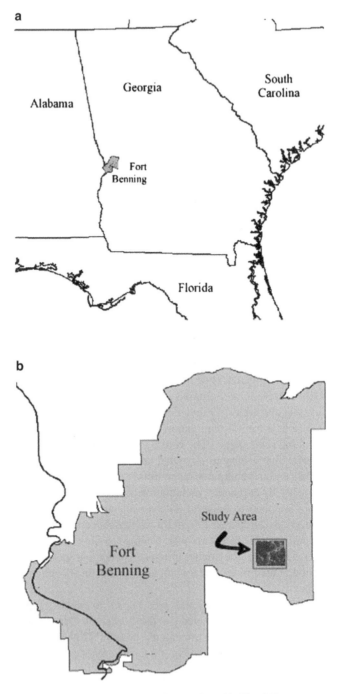

Fig. 6.2 Location of Fort Benning study area (inset enlarged in Fig. 6.3)

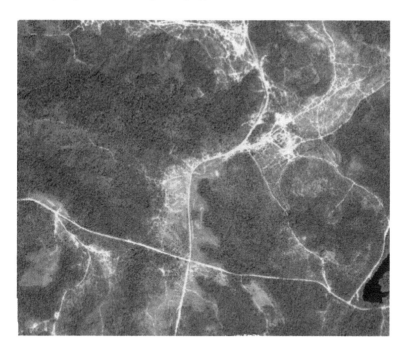

Fig. 6.3 Study area. Note light-colored dirt roads and trails used for soldier training, lighter areas that are sparsely vegetated uplands, and darker forested lowlands

tortoises and has historically been used for training by the Army. This use is likely to increase over the next decade as tank maneuver training is added to the Fort Benning mission. The area is largely composed of forest, which primarily occupies lower elevations, while dirt roads and trails share the uplands with gopher tortoises.

6.3.2 State Variables and Scales

The model, SimGT, uses two agents: tortoises and patches. The first represents individual tortoises that hatch, live, grow, reproduce, move, and die on a regular square grid of 60×60 m patches via 1-month model time steps. The patches are defined by their vegetation coverages (herbaceous and woody), number of eggs laid, fire frequency and size, tortoise capacity, number of initial burrows, and soil type. Many of these variables are dynamic during a simulation. Tortoises are dynamically defined by their location, carapace length, sex, age, energy reserves, move-motivation, distance moved, and whether they are a "disperser." In addition, the following variables—user-settable with slider controls—are used to guide the simulation (Table 6.1).

Table 6.1 User-settable variables

Name	Default value	Description
Juvenile-CL	50 mm	The initial carapace size for new juveniles
Carapace-rate	20 mm/year	The carapace growth per year
Max-CL-female	330 mm	The maximum carapace size for females
Max-CL-male	310 mm	The maximum carapace size for males
Min-juvenile-age	1 year	The age at which hatchlings mature to juveniles
Min-adult-age	10 year	The age at which juveniles mature to adults
Min-senior-age	40 year	The age at which adults mature to seniors
Max-GT-per-HA	3.0 GT	The maximum carrying capacity
Initial-tortoises	100	The number of initial tortoises randomly distributed
Dispersers	2%	The percent of tortoises that disperse
Egg-to-juvenile-survival	5%	Percent of eggs that survive through hatchlings to juveniles
Juvenile-death-prob	5%/year	Annual probability of death for juveniles
Adult-death-prob	3%/year	Annual probability of death for adults
Senior-death-prob	20%/year	Annual probability of death for seniors
Female-reproduction-prob	60%/year	Annual probability of each female laying eggs
Eggs-per-female	6/year	Number of eggs laid by each reproducing female
Veg-growth?	T/F	A toggle for vegetation growth
Woody-growth-rate	5%/year	Growth rate for woody vegetation (trees)
Herb-growth-rate	10%/year	Growth rate for herbaceous vegetation
Fire?	T/F	A toggle for lightning-induced fire
Lightning	0.01 strikes/year/ha	Number of lightning strikes
Burn-probability	20%	Probability that a patch will burn
Pct-herbs-lost-to-fire	10%	Percent of herbaceous vegetation destroyed in a patch by fire
Pct-woody-lost-to-fire	95%	Percent of wood vegetation destroyed in a patch by fire

6.3.3   Process Overview and Scheduling

This model runs with a 1-month time step, accomplishing the following steps:

1. Vegetation grows (optionally, May only).
2. Wildfires burn (optionally, May only).
3. Hatchlings are promoted to juveniles (April only).
4. Smallest tortoises migrate if the temperature is greater than 50°F and if the area is over carrying capacity.
5. A small randomly selected set of tortoises disperse if the temperature is greater than 50°F.
6. All tortoises age 1 month and grow.

7. All tortoises increase energy reserves if the temperature is greater than 50°F.
8. All females of an appropriate age reproduce (April only).
9. Some tortoises die based on energy levels and age.

6.3.4   Design Concepts

6.3.4.1   Emergence

By capturing the behavior of many individuals over time and space, we anticipate the illumination of an emergent behavior for populations and metapopulations.

6.3.4.2   Migration

Patches that are overpopulated result in the migration of younger individuals until carrying capacity is reached. This primarily means that the youngest individuals are forced to the edges of habitable areas.

6.3.4.3   Dispersion

Regardless of tortoise density and tortoise age, some tortoises have been documented to simply disperse (Diemer 1992; Wilson et al. 1994). Based on the literature, a small percentage of tortoises are selected at random to move at randomly large distances.

6.3.4.4   Aging and Reproduction

Over time tortoises age and reproduce. This process, at the scale of the individual, results in the emergence of population age cohorts.

6.3.4.5   Death

As tortoises age from eggs to seniors, their probability of death changes. Few, of course, are lucky enough to reach old age.

6.3.5   Initialization

Model initialization involves user-settable variables (as listed previously in Sect. 6.3.2), reading maps to set patch variables (discussed below in "input map

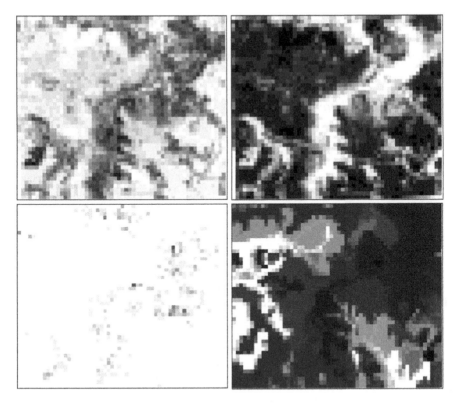

Fig. 6.4 Habitat suitability maps for gopher tortoises generated based on herbaceous density (*top left*), woody density (*top right*), burrow density (*lower left*), and soil suitability (*lower right*)

development"), and using a few hard-coded values that capture minimum and maximum average temperatures and changes in carrying capacity by month. The temperatures can easily be modified in the NetLogo procedure view by editing code in the "set-location" procedure. Temperatures are set for the Fort Benning area based on historic weather measurements. The relative carrying capacity adjustments are set to 1.0 for winter months and up to 3.0 for summer months to capture the notion that carrying capacity densities vary over the year based on food availability. To use this model for another area, these values will need to be reset, and the input maps will need to be recreated (Fig. 6.4).

6.3.6 Input

The interface (Fig. 6.5) enables users to change operation parameters and to view the state of the system during simulation. The map at the top right provides a 55×45

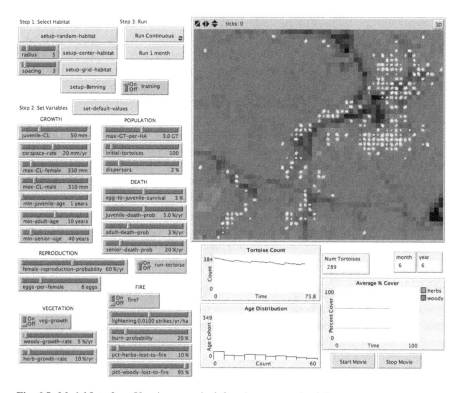

Fig. 6.5 Model Interface. User inputs to the left and outputs to the right

raster geographic information system (GIS)-type view of the system during simulations. Model output displays patches of woody vegetation as different shades of red, based on the percent cover. Similarly, areas of differing shades of green represent the percent of herbaceous vegetation within each patch. Figure 6.5 is a black and white illustration of these patches (black is no vegetation). Tortoises are represented as tortoise icons colored pink for females and blue for males, and size is based on their carapace size (which is based on age). During simulations the patches and tortoises can be probed to allow tracking of their individual states over time. Buttons on the top left of the interface can be clicked to initialize and run SimGT. Model parameters can be changed with the many sliders along the lower left side of Fig. 6.5. Various plots under the main map provide summary feedback during model runs to show the total population count, the age distribution, vegetation covers, and time (month and year).

The spatial habitat in this model can be initialized using input from digital maps based on a GIS or with software-based descriptions of simplified habitats. In this case, we used GRASS[2] GIS software to develop each of the model's five input maps. These GIS maps provide the required landscape description for the area of interest: burrows, soil, trails, percent herbaceous growth, and percent woody growth. The burrows map holds the total number of burrows that exist in each 60×60 m grid cell. The percent woody and herbaceous map uses integer values between 0 and 100 to capture the percent of the land in each cell covered by woody (tree and shrub) and herbaceous (forbs and grasses) vegetation. This vegetation density map was created by processing aerial imagery with a resolution of 1 m using the unsupervised maximum likelihood classifier i.cluster and the classification program i.maxlik. The resulting map had 49 categories, which were visually matched to four landcover types: water, urban, forest, and four levels of herbaceous density. From this, a forest map was generated with a value of 100% assigned to cells containing forest and a value of 0 to those that did not. These were resampled to a 60 m cell resolution to generate the average forest cover. Similarly, the levels of herbaceous density were assigned values of 25, 50, 75, and 100% and then averaged over 60 m cells. Likewise, the trails map provides an integer value between 0 and 100 representing the percent of land covered by roads and trails.

Gopher tortoises prefer upland habitat types (Landers 1980; Auffenberg and Franz 1982; Diemer 1986) and they seem to respond more to physical conditions than to plant associations (Campbell and Christman 1982). The most important aspect of the land itself is its ability to support the construction of burrows.

The soils suitability map was constructed as an index map, with values ranging from 0 to 100 used to indicate the value of each cell's soils for supporting tortoise burrow development. This map was developed as a linear regression applied to the model's independent variables for each patch, i.e., amount of a particular soil type passing through a 200 mesh (200 wires/in.) sieve, slope, slope aspect, and watershed area that drains through the patch. The dependent variables are known locations of historic tortoise burrows and a random sampling of an equal number of locations in the study area that represent areas where no burrows are found. A linear regression equation was obtained using MacAnova software (Oehlert and Bingham 1997) and applied to every location across the study area using the GRASS r.mapcalc program at a resolution of 1 m. The result was then averaged across the 60 m resolution map. The four resulting maps (Fig. 6.6) were written into ASCII files that could then be read into the NetLogo modeling environment. A grey scale is used in each of the images to show relative habitat suitability. Darker gray tones in the map correspond to higher suitability value. Gray scales represent index values ranging from 0 to 100 except in the burrow density map (lower left), which uses a data range of 0–8 burrows/60-m cell. The tortoise burrow density map is simply a

---

[2] GRASS, originally an acronym for the Geographic Resources Analysis Support System, is an open source GIS developed starting in 1982 by the U.S. Army Construction Engineering Research Laboratory (USACERL), the predecessor organization of the U.S. Army Engineer Research and Development Center–Construction Engineering Research Laboratory (ERDC-CERL). GRASS is currently developed and maintained by the Open Source Geospatial Foundation (http://www.osgeo.org/).

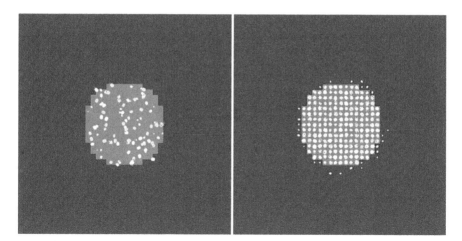

Fig. 6.6 Maps developed by GRASS r.mapcalc program using linear regression equation applied to every location across the study area

Table 6.2 Tortoise state variables

The length of the carapace in mm
F or M
In years, but incremented by 1/12 each month
Energy units used to limit dispersal efforts
A desire-to-move index (0–1)
Counts of the number of cells traversed
A desire to move longer distances (T/F)

count of the total number of burrows in each 60-m cell from a database developed in the 1990s (unpublished work performed for Fort Benning by Auburn University, Auburn, AL).

6.3.7   Submodels

6.3.7.1   Tortoise Development

In the model, each tortoise is defined by a set of state variables that change over the simulation time (Table 6.2) and a set of constants that can be modified by the modeler.

Because tortoises have a very long lifespan, it is necessary to model a population over the course of a century, at least. This way it is possible to capture population dynamics across multiple generations. A time step of 1 month was chosen to capture the individual behaviors that affect a population's viability over this time period.

The spatial resolution was chosen to ensure that the vast majority of tortoise movements do not go beyond neighboring areas in any given time step. Adult gopher tortoises are reported to be active only 9.2% of the time (Auffenberg and Iverson 1979).

Table 6.3 Initial population parameterization

Maximum of tortoises/ha	3.0
Initial number of tortoises	100

Table 6.4 Weather parameters

Average monthly high temperature	A temperature for @ month
Average monthly low temperature	A temperature for @ month
Relative density adjust	A% value for @ month

Over the course of a year, tortoises were observed to be above ground only 0.8% of the time (Eubanks et al. 2003). Various studies have looked at home ranges for tortoises during a year: 0.0002 and 1.435 ha with a mean of 0.492 ha (Mitchell 2005), 365 ± 265 m^2 (Butler et al. 1995), and 0.88 ha for males/0.31 ha for females (Diemer 1992). Home range size appears to be inversely proportional to the density of herbaceous ground cover (Auffenberg and Iverson 1979) within suitable gopher tortoise habitat. The mean home range size for subadults is 0.05 and 0.01 ha for juveniles (Diemer 1992). Based on this information, landscapes for the model were raster gridded with a resolution of 60 m, creating a regular array of 0.36 ha patches. This provided a home range area that could naturally support two to three tortoises and captures typical tortoise movement.

An initial population of tortoises is distributed across the simulation space. Using the Fort Benning study area, each cell that contained one or more burrows is assigned a single tortoise, who is given an age between 1 and 50 and a sex at random. The carapace length is set based on the assigned age. For graphic display purposes, a tortoise icon is used. Females are set to pink and males to blue. The icon size is set based on carapace size. When the model is initialized for a randomized landscape, a user-specified number of tortoises are created and then are randomly assigned locations in the space until each tortoise is in an area with herbaceous cover greater than 30%. The maximum number of tortoises that can exist in the best habitat is also set, but can be changed through the interface. These variables are listed in Table 6.3.

Gopher tortoises are warm-weather creatures that remain in burrows during the months of colder weather (Butler et al. 1995). Activity in the model for a given month is based on the monthly average low and average high temperatures. Outside the burrow, activity only happens when the average of these temperatures are above 25°C.

A set of values representing monthly average conditions is stored in the model and can be easily changed to help localize the model. Lists contain average daily low and high temperatures and a relative tortoise density adjustment (Table 6.4). This information is then used for habitat suitability and tortoise activity calculations.

6.3.7.2 Model Simulation Dynamics

The model uses 1 month as its time step. At each increment, several general events may occur. First, the optional vegetation growth and landscape fires are processed— these dynamics can be turned on and off through the main interface. Then, the patch state variables are updated including the calculation of the maximum number of gopher tortoises allowed. Next, incubating gopher tortoise eggs are transitioned into hatchlings. Finally, the gopher tortoises are allowed to migrate, grow, eat, reproduce, and die. Each of these steps is discussed below.

6.3.7.3   Growth of Vegetation

Gopher tortoises need ground-level access to herbaceous forage, including broad-leaved grasses, wiregrass, legumes, and other plant materials (Cox et al. 1987). In otherwise suitable habitat, it is the lack of sufficient vegetation that will drive a population out of an area. As herbaceous cover is lost through succession to a woody climax forest, home ranges expand, tortoise densities drop, and eventually tortoises cannot be supported any longer.

To accommodate vegetation growth, woody and herbaceous vegetation grows according to the following logistic growth equation at each time step:

$$P = P + rP(1 - P / K)$$

where P is the population expressed in percent cover, r is the maximum growth rate, and K is the maximum cover (100%).

Woody growth after a density of 80% limits the maximum herbaceous growth linearly until it equals 0; then, woody growth is 100%. Vegetation growth can be turned on or off depending on experimental needs. If turned on, and without any process decreasing vegetation cover, the entire area turns into an uninhabitable forest.

Fires in the longleaf pine ecosystem of the southeastern USA have been an important component in maintaining the gopher tortoise habitat. Both naturally occurring fires and prescribed burning serve to eliminate woody vegetation and allow the herbaceous vegetation to dominate when it occurs with sufficient frequency. In many areas, however, fires are routinely suppressed to protect life and property. As a result, an unnatural overdevelopment of the forest canopy occurs, and gopher tortoise habitat quality suffers. In a study of areas subject to prescribed burning compared with others existing under a policy of fire suppression, tortoise densities were measured to be three times greater in the burned areas (Landers and Buckner 1981) than the unburned. We sought to emulate these fire characteristics in the model.

In the model user interface, four fire-related variables can be set. The lightning-strikes-per-year-per-hectare variable is used to set the random frequency of fire starts and a burn-probability value establishes the potential for fire to spread from cell to cell. Then, there are two variables that set the percent herbaceous and woody vegetation cover that is lost to each fire. Vegetation growth and fires occur once every 12 months and establish ever-changing random patches that mainly remove woody vegetation.

The simplest assumption to model is that the current landscape pattern of vegetation will be maintained through human activities. Optionally, vegetation growth could be turned on and vegetation-control efforts could be scheduled and simulated.

6.3.7.4   Set Tortoise Carrying Capacity

The next step in the monthly simulation is to establish the current tortoise carrying capacity for each cell on the simulated landscape. For this model, each tortoise is associated with a home-range centroid that is attached to a patch. Behavior rules allow each tortoise to relocate that centroid to a neighboring patch each month. The attractiveness of each patch is computed based on vegetation density, tortoise density, and seasonality (which adjusts the food value of the vegetation). The attractiveness is computed as a maximum density value. If the 0.36 ha patch has more tortoises than the maximum, tortoises leave, beginning with the youngest/smallest, until the maximum density is reached. This only occurs during months when tortoises are active.

In the model the number of tortoises that could be added to each cell is calculated as follows. The user-set maximum-tortoise-per-acre variable (Table 6.3) is multiplied times the monthly relative-density-adjust value (Table 6.4) and by a vegetation index based on the percent herbaceous cover. The last value ranges between 0 (for a percent herbaceous cover ≤20%) and 1 (for a percent cover ≥80%). This value times the number of hectares per cell gives the current tortoise carrying capacity. The difference between that value and the current tortoise population gives the number of tortoises that could be added to the cell.

6.3.7.5   Process Eggs and Hatchlings

The process of egg production and survival through the hatchling year all takes place in a relatively small area. For this model, the most important factor is the survival of eggs to the juvenile stage. That is reported as 94.2% (Alford 1980) and 92.3% (Witz et al. 1991). Hatchlings range from 1.8 to 24.2 m away from the nest with an average distance of 8.3 m (McRae et al. 1981). Therefore, eggs and hatchlings are modeled as a whole with progression to the juvenile stage occurring after a year, based on an overall survival rate that can be set by the user. The number of eggs and hatchlings is stored as a variable associated with patches; they are not treated as individual entities, which is possible because they do not move out of their 0.36 ha patch during that time. In month 4 (April), those that survive are initialized as model agents (i.e., mobile individual tortoises).

6.3.7.6   Tortoise Movement

Gopher tortoises prefer to remain close to their burrows. One study discovered that female feeding takes place within 17 m of the current burrow and the average radius of feeding areas is 11.9 m (Smith 1995). In another study it was found that 95% of all feeding activity took place within 30 m of the active burrow with an average of only 13 m (McRae et al. 1981).

Tortoise motivation to relocate is not well understood, but some clues have been published. Home range sizes vary considerably. For example, Smith measured ranges from 0.002 to 1.435 ha (Smith 1995) over 500 days. Tortoises relocate even from suitable habitat, moving over 0.45 ha in favorable conditions (McRae et al. 1981). Additionally, there may be movement from xerophytic summer areas to mesophytic lowlands in winter (McRae et al. 1981; Means 1982; Breininger et al. 1988). It has also been suggested that home range size may double or triple in the late summer and fall compared with spring sizes (Auffenberg and Iverson 1979; McRae et al. 1981). This increase may be based on variation in seasonal vegetation density, which is lower in summer and fall than in spring; and seasonal migration to and from overwintering areas (McRae et al. 1981).

In this model, tortoises can move to neighboring cells if the average low temperature for that month is greater than 50°F. Dispersal only happens when the total number of tortoises in a cell is greater than the previously calculated capacity for the cell for that month. Required dispersal begins with the smallest tortoises and continues until no patch is overpopulated.

6.3.7.7   Tortoise Migration

For the purposes of this model, dispersing tortoises are those that move just far enough to avoid overpopulation and migrating tortoises are those that travel farther than needed to avoid overpopulation. There is little information in the literature reporting the migration success rates, but there is strong evidence for the occasional migrating tortoise. In the Eubanks et al. study (2003), two surviving tortoises were considered emigrants from the study site. Both were males and moved significant distances (4.8 and 6.4 km) from the study site before settling. In another study a migrating subadult was tracked 0.74 km (Diemer 1992). None of the authors report any particular motivations for these dramatic moves in an otherwise sessile population. To accommodate for these findings in the model, some percentages of adult tortoises each year (e.g., 2%) are given an extra motivation to migrate. For the active periods of the year, assuming they have enough energy stored, they move up to 800 m.

6.3.7.8   Tortoise Growth

The model provides a user selected eggs per female variable and a frequency of female egg laying variable. In the model, these are applied to each female based on carapace length, which is based on age as follows. The CL is set to an initial value for a yearling juvenile and increases each year according to the following equation:

$$C_{t+1} = C_t + R(1 - C_t / C_{max})$$

where C is the carapace length and R is the initial rate of growth.

The initial CL, initial CL growth rate, and maximum CL are user settable variables that default to 10, 20, and 380 cm, respectively. The probability of a female nesting in any given year and the number of eggs per nest are also user settable with default values of 60% and 6 eggs respectively.

Table 6.5 Tortoise growth parameters

Minimum carapace length of juveniles	50 mm
Growth rate	20 mm/year
Maximum carapace size for females	330 mm
Maximum carapace size for males	310 mm
Age at which a hatchling becomes a juvenile	1 year

6.3.7.9 Tortoise Feeding

As a poikilotherm, the gopher tortoise has very low caloric requirements, but if caught in a developing deep forest it can conceivably starve to death, or succumb to disease when weak. No energetics information has been located in the literature that describes the caloric cost of various behaviors (e.g., foraging, burrow digging, traveling) and the caloric benefit of feeding. Unfortunately, without caloric costs and benefits of various activities, there is little hard data to help build a PVA. Because tortoises can ingest so many different materials, our modeling effort assumes that (1) tortoises will not starve, (2) tortoises caught in unsuitable habitat will continue moving until suitable habitat is found, and (3) the movement of younger/smaller tortoises from overpopulated areas will capture the food competition activity. In the model, every month in which the average low temperature is above 50°F, each tortoise's energy reserve level is increased by 0.5, up to a maximum of 5.0. (These numbers represent unitless values on a scale of 0–5 that indicate the size of a tortoise's energy reserves.) (Table 6.5).

6.3.7.10 Tortoise Reproduction

Gopher tortoises reach sexual maturity relatively late. For the purposes of this model, 10 years was used as the age at which tortoises begin to reproduce. Once mature, males compete for females. Larger males tend to dominate the breeding game and fertilize the majority of clutches by winning aggressive encounters with smaller males (Moon et al. 2006). They also mate with several females (Douglass 1976; Epperson 2003). Clutches generated by larger females tend to be sired by a single male, while smaller females have clutches sired by multiple males (Moon et al. 2006). In one study, 28.6% of clutches showed multiple paternity (Colson-Moon 2003). Breeding can begin as early as February (Dietlein and Franz 1979) and may extend into September.

Perhaps the most thoroughly studied aspect of the gopher tortoise is the number of eggs laid in a nest. Reported mean clutch sizes are 4.8 (Yager et al. 2006), about 6 (Diemer 1986), 3–11 (Dietlein and Franz 1979), 9 in southern Florida (Iverson 1980), 4.8 at Camp Shelby (Epperson 2003), 5.76 (Smith 1995), 6–7 in southwest Georgia (Landers 1980), 4–6 in north central Florida (Iverson 1980; Diemer 1986), and 5.4 in Duval County (Butler et al. 1996). In general, the number of eggs laid seems to vary across regions and states, but is associated with carapace length

Table 6.6 Tortoise reproduction parameters

Parameter	Value
Age at which a tortoise becomes reproductive	10 years
Age at which a tortoise is no longer productive	40 years
Eggs laid per female	6
Probability of female egg-laying	60%/year

Table 6.7 Tortoise survival parameters

Minimum juvenile age	1
Minimum adult age	10
Minimum senior age	40
Egg-to-juvenile survival	5%
Probability of juvenile death	5%/year
Probability of adult death	1%/year
Probability of senior death	20%/year

(Iverson 1980; Landers et al. 1980; Diemer and Moore 1994; Rostal and Douglas 2002; Colson-Moon 2003).

Fifty-five percent of females were gravid each year in Central Florida (Colson-Moon 2003) and elsewhere it was reported that females lay eggs 2 out of every 3 years (Lohoefener and Lohmeier 1981). In the model, 60% of female tortoises lay eggs each year.

The parameters listed in Table 6.6 are used to add to the number of eggs/juveniles in each patch as follows. In month 4 (April) females between the minimum and maximum reproduction ages are nominated to lay eggs. A random number is generated for each female between 0 and 100, and those with a number less than the egg-laying probability lay eggs at the egg-laying rate.

6.3.7.11   Tortoise Death

Tortoises may die because they have exhausted energy reserves or they are randomly unlucky based on parameters listed in Table 6.7. The probabilities of death are adjusted from annual values to monthly values and applied each month.

6.4   Simulation Experiments

6.4.1   Simple Circular Habitat Experiment

An expectation of individual-based modeling is that behaviors and responses of populations will emerge. This first experiment seeks to compare model outputs with snapshot measurements in the field of populations that have survived and potentially stabilized in relatively consistent environmental conditions. In this experiment,

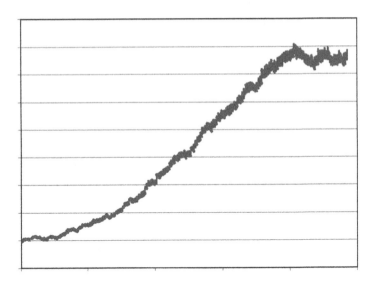

Fig. 6.7 Population logistic growth curve

there is a central 163 ha circle of excellent habitat (80% herbaceous and 0% woody vegetation) surrounded by forest (100% woody and 0% herbaceous).

Wildfires and vegetation growth are turned off for this simulation. A view of the simulation at initialization is displayed in the left image in Fig. 6.6. The herbaceous area is the green center section and the surrounding forest is in red. All initial tortoises are randomly located in the herbaceous area. By the 163rd year of the simulation, the maximum density is reached and the young tortoises are being displaced into the surrounding forest (right image in Fig. 6.6). There, the individual tortoises die after several months as a result of being forced to exist without food. This completes the feedback loop that results in a population logistic growth curve (Fig. 6.7). Note that the population in this experiment overshoots an overall carrying capacity and oscillates around about 750 adults and juveniles.

6.4.2 Model Application

The SimGT model was initialized with prepared maps and resulted in the image in Fig. 6.8. Herbaceous and woody index values were used as vegetation (percentage) densities. The cells measure 60×60 m, resulting in a monthly home range of 0.36 ha. The amount of woody vegetation is represented by intensities of red ranging from black to bright red, and herbaceous vegetation is similarly represented in shades of green. The area is populated with gopher tortoises at a density of one per burrow, and the tortoises are assigned random ages and genders. The model parameters remain set to those discussed in Sect. 6.3.2. Vegetation growth and random wild fires are turned off to represent land management that will maintain the status quo.

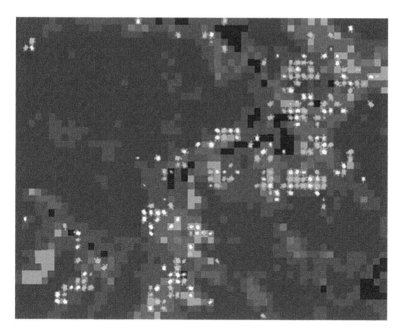

Fig. 6.8 A snapshot of the model display showing darker male tortoises (blue when running the model) and lighter female tortoises (pink in the model). The large contiguous darker areas are higher-density trees. Tortoises are clustered in open areas with low tree cover. When running the model, patches are colored in darker greens with increased tree density and darker orange with increased herbaceous density

For this model, the "world" ends at the boundaries and does not wrap east–west and north–south as is often done to avoid edge effects.

The model was run 100 times, each for a 100-year simulation (Fig. 6.9). Note that the vegetation did not change over the simulation time (no growth or loss by fire). In all cases the overall population persisted in the area, but at the end of the 100 years different subareas experienced loss of tortoises, while others developed a local population. Sample 100-year simulations of this model are presented in Fig. 6.9. The resulting patterns of gopher tortoise settlement are presented for each of three consecutive simulation runs (top two and bottom left images). Note that these patterns can be quite different—simply because of the random movement of migrating individuals. Areas B, D, and F show consistent expectation of persistence of tortoises and may be categorized as source or refugia areas. Areas A, C, and E range from few/no individuals to significant populations during different simulations. Semi-isolated areas like G show no settlement after 100 years in these simulations, but occasionally they can be colonized. With nearly identical starting points (only the age and sex of initial individuals are different) and identically constant boundary conditions, the spatial distribution of the population can be substantially different. Such results match field experiences of inexplicable changes in population patterns of distribution. The bottom right image in Fig. 6.9 summarizes the results of 100 runs of the model and indicates the average number of tortoises

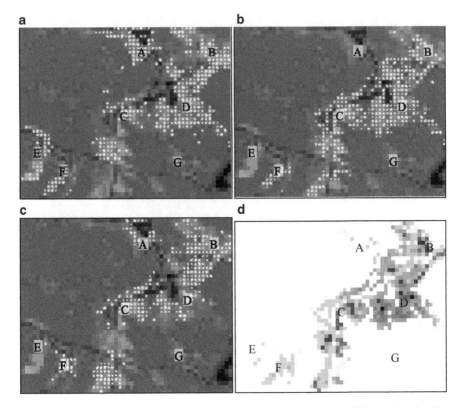

Fig. 6.9 Sample 100-year simulations. Individual simulation results after 100 years (**a–c**) and a summary of 100 runs (**d**), where the darker the cell, the more frequent the cell contained one or more tortoises at simulation end

occurring in each location after 100 years. The results range from an average of 0 tortoises (white) to 1.44 tortoises (black).

This analysis could assist in land management decisions. For example, areas B, C, and D clearly remain the best areas for ensuring a viable population. Areas A, E, F, and G provide potential for enhancing the population, but do not ensure natural population persistence as modeled. Future losses of tortoises in these areas should be viewed as natural, and the areas could be recolonized by tortoises from the B, C, and/or D core areas.

6.5 Discussion

This work investigated whether a spatially explicit population viability model can be built for a species whose population dynamics have been studied very little but the behavior of individuals has been studied in detail. Much of the literature falls into two categories: short-term studies on individual behaviors and snapshots of population structure and distribution. We have shown that there is sufficient published information about the gopher tortoise to develop a spatially explicit individual-based simulation model that could help guide the development of land management plans. In this model, tortoise agents are programmed with behaviors that move them among adjacent home ranges based on habitat suitability, average monthly air temperature, and tortoise density. To allow natural settlement of suitable habitat, unoccupied areas must be virtually contiguous with already populated areas. This contiguity must be maintained over decades with appropriate management to ensure the duplication of the historic effects of wildfires. This information and the model can be used to compare the utility of alternative tortoise reserve plans.

The current model contains only a rudimentary representation of the effect of military training activities on tortoise densities. More published research on the effects of training—and associated maps of training intensity and types—would help the model respond appropriately to military land management alternatives. Also, more research is required to understand the motivations and strategies of tortoise dispersal and settlement—i.e., how far a tortoise can and will travel when dispersing. Another needed area of research is gopher tortoise dispersal triggers. We do not know how permeable different habitat types are to tortoises or what strategies tortoises use to optimize the probability of finding suitable habitat.

6.6 Conclusion

Models are abstractions of the real world and are best developed to answer specific questions. Much of the cited gopher tortoise literature describes the behavior of individual tortoises or demographic characteristics of tortoise populations. This study demonstrated the utility of this information in the development of a comprehensive description of tortoise behavior in the form of a spatially explicit simulation model that allows forecasts of the long-term viability of tortoise populations.

SimGT could easily be applied to other areas at or around Fort Benning, as well as other areas in the southeastern USA by developing the required input maps. The model will remain available to any interested parties as open source by contacting the authors.

Acknowledgment This project was supported with funding through ERDC-CERL Project 140644, "Habitat-Centric SAR (Species at Risk) Research—Multi-species PVA (Population Viability Analysis)."

References

Akçakaya HR (2002) RAMAS GIS: linking spatial data with population viability analysis (version 4.0). Applied Biomathematics, Setauket

Alford RA (1980) Population structure of *Gopherus polyphemus* in northern Florida. J Herpetol 14(2):177–182

Auffenberg W, Franz R (1982) The status and distribution of the gopher tortoise (*Gopherus polyphemus*). In: Bury RB (ed) North American tortoises: conservation and ecology Wildlife Research Report No. 12, US Fish and Wildlife Service, Washington, pp 95–126

Auffenberg W, Iverson JB (1979) Demography of terrestrial turtles. In: Harless M, Norlock N (eds) Turtles: research and perspectives. Wiley-International, New York, p 718

Breininger DR, Schmalzer PA, Rydine DA, Hinkle CR (1988) Burrow and habitat relationships of the gopher tortoise in coastal scrub and slash pine flatwoods on Merritt Island, Florida. Florida Game and Fresh Water Fish Commission, Nongame Wildlife Project Report GFC-84-016, pp 1–238

Butler JA, Bowman RD, Hull TW, Sowell S (1995) Movements and home range of hatchling and yearling gopher tortoises, *Gopherus polyphemus*. Chelonian Conserv Biol 1(3):173–180

Butler JA, Bowman RD, Hull TW (1996) Reproduction of the tortoise (*Gopherus polyphemus*) in northern Florida. J Herpetol 30(1):14–18

Campbell HW, Christman SP (1982). The herpetological components of Florida sandhill and sand pine scrub associations. In: Scott NJ Jr (ed) Herpetological communities. Wildlife Research Report 13, U.S. Fish and Wildlife Service, pp 163–171

Colson-Moon JC (2003) Reproductive characteristics, multiple paternity and mating system in a Central Florida population of the gopher tortoise, *Gopherus polyphemus*. University of South Florida, Tampa

Cox J, Inkley D, Kautz R (1987) Ecology and habitat protection needs of gopher tortoise (*Gopherus polyphemus*) populations found on lands slated for large-scale development in Florida. Nongame Wild Program Technical Report No. 4, Florida Game and Fresh Water Fish commission, Office of Environmental Services, Tallahassee

Rostal DC, Jones DN Jr (2002) Population biology of the gopher tortoise (Gopherus polyphemus) in southeast Georgia. Chelonian Conservation and Biology 4: 479–487

Diemer JE (1986) The ecology and management of the gopher tortoise in the southeastern United States. *Herpetologica* 42(1):125–133

Diemer JE (1992) Home range and movement patterns of the tortoise *Gopherus polyphemus* in northern Florida. J Herpetol 26(2):158–165

Diemer JE, Moore CT (1994) Reproduction of gopher tortoises in north-central Florida. In: Bury RB, Germano DJ (eds) Biology of North American tortoises. Fish and Wildlife Research 13, U.S. Fish and Wildlife Service, pp 129–137

Dietlein NE, Franz R (1979) Status and habits of *Gopherus polyphemus*. In: Amant E (ed) Proceedings of Symposium of Desert Tortoise Council, Tucson, pp 175–180

Douglass JF (1976) The mating system of the gopher tortoise, *Gopherus polyphemus*, in southern Florida. University of South Florida, Tampa

Epperson DM (2003) Impacts of a non-native species, *Solenopsis invicta* (red imported fire ants), on a keystone vertebrate, *Gopherus polyphemus* (gopher tortoise), and its associated commensal fauna. Ph.D. Dissertation, Clemson University

Eubanks JO, Michener WK, Guyer C (2003) Patterns of movement and burrow use in a population of gopher tortoises (*Gopherus polyphemus*). Herpetologica 59(3):311–321

Guertin, Patrick J (2005) Training Restrictions on Army Lands Due to High Priority Endangered Species. ERDC/CERL TR-05-12. Champaign, IL: US Army Engineer Research and Development Center, Construction Engineering Research Laboratory (ERDC-CERL)

Iverson JB (1980) The reproductive biology of *Gopherus polyphemus* (Chelonia: Testudinidae). Am Midl Nat 103(2):353–359

Landers JL (1980) Recent research on the gopher tortoise and its implications. In: Franz R, Bryant RJ (eds) The dilemma of the gopher tortoise—is there a solution? Proceedings of the 1st Annual Meeting, Gopher Tortoise Council. Florida State Museum: Gainsville, pp 8–14

Landers JL, Buckner JL (1981) The gopher tortoise: effects of forest management and critical aspects of its ecology. Forest Productivity and Research: Technical Note No. 56. Bainbridge: International Paper Company, Wood Products and Resources Group, Southlands Experiment Forest. p 7

Landers JL, Garner JA, McRae WA (1980) Reproduction of gopher tortoises (*Gopherus polyphemus*) in southwestern Georgia. Herpetologica 36(4):353–361

Lindenmayer DB, Burgman MA, Akçakaya HR, Lacy RC, Possingham HP (1995) A review of the generic computer programs ALEX, RAMAS/space and VORTEX for modelling the viability of wildlife metapopulations. Ecol Model 82(2):161–174

Lohoefener R, Lohmeier L (1981) Comparison of gopher tortoise (*Gopherus polyphemus*) habitats in young slash pine and old longleaf pine areas of southern Mississippi. J Herpetol 15(2):239–242

McRae AW, Landers JL, Garner JA (1981) Movement patterns and home range of the gopher tortoise. Am Midl Nat 106(1):165–179

Means DB (1982) Responses to winter burrow flooding of the gopher tortoise (*Gopherus polyphemus* Daudin). Herpetologica 38(4):521–525

Mitchell M (2005) Home range, reproduction, and habitat characteristics of the female gopher tortoise (*Gopherus polyphemus*) in southeast Georgia. M.S. thesis, Georgia Southern University, Statesboro. http://eaglescholar.georgiasouthern.edu:8080/jspui/bitstream/10518/1588/3/Mitchell_Maggie_J_200508_MS.pdf

Moon JC, McCoy ED, Mushinsky HR, Karl SA (2006) Multiple paternity and breeding system in the gopher tortoise, *Gopherus polyphemus*. J Hered 97(2):150–157

Mushinsky HR, McCoy ED (1994) Comparison of Gopher tortoise populations on islands and on the mainland in Florida. In: Bury RB, Germano DJ (eds) Biology of North American tortoises fish and Wildlife Research Report 13, U.S. Fish and Wildlife Service, Washington, pp 39–48

Oehlert GW, Bingham C (1997) MacAnova user's guide. University of Minnesota, School of Statistics, St. Paul

Rostal DC and Douglas N. Jones, Jr. (2002) Population biology of the gopher tortoise (Gopherus polyphemus) in southeast Georgia. Chelonian Conservation and Biology 4: 479–487

Smith LL (1995) Nesting ecology, female home range and activity, and population size-class structure of the gopher tortoise, *gopherus polyphemus*, on the Katherine Ordway Preserve, Putnam County Florida. Bull Fla Mus Nat Hist 38:97–126

Wilensky U (1999) NetLogo. Computer software. Center for Connected Learning and Computer-Based Modeling, Northwestern University, Evanston. http://ccl.northwestern.edu/netlogo/

Wilson DS, Mushinsky HR, McCoy ER (1994) Home range, activity, and use of burrows of juvenile gopher tortoises in Central Florida. In: Bury RB, Germano DJ (eds) Biology of North American tortoises. Fish and Wildlife Research Report 13, National Biological Survey, Washington, pp 147–160

Witz BW, Wilson DS, Palmer MD (1991) Distribution of *Gopherus polyphemus* and its vertebrate symbionts in three burrow categories. Am Midl Nat 126:152–156

Yager L, Hinderliter M, Balbach H (2006) Response of gopher tortoises to habitat manipulation by prescribed burning. Construction Engineering Research Laboratory, Champaign

Chapter 7
Using Demographic Sensitivity Testing to Guide Management of Gopher Tortoises at Fort Stewart, Georgia: A Comparison of Individual-Based Modeling and Population Viability Analysis Approaches

Tracey D. Tuberville, Kimberly M. Andrews, James D. Westervelt, Harold E. Balbach, John Macey, and Larry Carlile

7.1 Background

One of the challenges with conserving rare species is identifying the most effective management targets; that is, the demographic traits most likely to positively influence population persistence through either manipulation of the habitat or the wildlife population. Furthermore, these targets should represent the most efficient use of limited resources, especially given that resource managers need to balance multiple, often complex issues (Reed et al. 2009). Population models can often aid managers in this process, and such models are frequently used to rank relative threats to specific populations, evaluate effects of proposed management actions or regulations, determine which demographic or ecological variables have greatest influence on extinction risk, and identify information gaps and research priorities (Tuberville et al. 2009 and references therein).

Population viability analysis (PVA) models represent a traditional modeling approach that has been used to support management decision-making for both game and nongame species. Unfortunately, robust PVA models require extensive population-level data for accurately estimating demographic parameters. Developing PVAs for rare species can be difficult, therefore, because complete life history

T.D. Tuberville (✉) • K.M. Andrews
Savannah River Ecology Lab, University of Georgia,
Drawer E, Aiken, SC 29802, USA
e-mail: tubervil@uga.edu

J.D. Westervelt • H.E. Balbach
Construction Engineering Research Laboratory, US Army Engineer Research
and Development Center, Champaign, IL, USA
e-mail: james.d.westervelt@usace.army.mil

J. Macey • L. Carlile
Fort Stewart Army Installation, Fish and Wildlife Branch,
1177 Frank Cochran Drive, Fort Stewart, GA 31314-4940, USA

J.D. Westervelt and G.L. Cohen (eds.), *Ecologist-Developed Spatially Explicit
Dynamic Landscape Models*, Modeling Dynamic Systems,
DOI 10.1007/978-1-4614-1257-1_7, © Springer Science+Business Media, LLC 2012

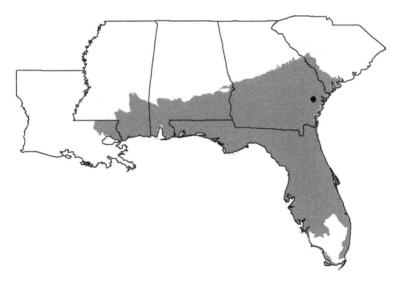

Fig. 7.1 Location of Fort Stewart Army installation (*dot*) within geographic range of the gopher tortoise (*shaded area*)

information and long-term population trend data often are not available. For many rare species, however, detailed information is known about their natural history and the behavior of individuals, including how they interact with each other and the landscape and how they respond to environmental cues. For these species, individual-based models (IBMs) may be more appropriate than PVA models for performing demographic sensitivity analysis; IBMs have the added advantage of imposing a spatially explicit landscape context.

The gopher tortoise (*Gopherus polyphemus*) is an example of a species whose life history data are incomplete, but whose natural history and individual behavior are well characterized. This species is considered to be declining throughout its range (Smith et al. 2006). It is federally listed as threatened in the western portion of its range (USFWS 1987) and is currently under consideration for listing throughout the remainder of its range (USFWS 2009). Gopher tortoise populations occur on many military installations throughout the southeastern USA (Wilson et al. 1997) and the species has been identified for management under the Army's Species at Risk (SAR) program. The SAR program seeks to develop proactive management strategies to ensure long-term viability of imperiled species that currently reside on military installations (NatureServe 2004).

Fort Stewart is the largest Army installation within the range of the gopher tortoise (Fig. 7.1). Given the current and anticipated increase in training demands at Fort Stewart—in terms of both intensity and spatial extent—resource managers are challenged to maintain viable populations of rare species within a limited or even diminishing footprint. One of the most practical ways to address this challenge is by improving demographic conditions for "at-risk" species through improvement of their existing habitat. Population models, used with demographic sensitivity analysis in particular, can help to determine the extent that habitat management alone can influence demographic parameters of rare species (such as the gopher tortoise) so that their abundance and likelihood of persistence will increase.

7.2   Objectives

As previously stated, IBMs and PVA models are both techniques that have been employed for developing long-term species management strategies. Although the two techniques have similar capabilities, some of the finer details can vary significantly. These variable details include the type of data required to develop the models, how scenarios are simulated, and the format and ways in which output data can be applied. A comparison of these two techniques can provide insight into which method is more sensitive to changes in parameter estimates and how a spatially explicit context may affect model results. Furthermore, as the SAR program and other management initiatives rely increasingly on predictive models, the use of multiple theoretical approaches can buffer the biases that are inherent in any one particular model. Eliminating biases will lend stronger support to management recommendations based on results from such models.

Our research objectives are listed here.

1. Develop a spatially explicit IBM to predict population-level dynamics that reflect the current understanding of the life history of the gopher tortoise. The IBM would simulate the collective behavior of individuals across multiple populations at Fort Stewart.
2. Conduct demographic sensitivity analysis of the developed IBM by comparing model output among simulations with different values for select demographic parameters.
3. Compare demographic sensitivity analysis results from the IBM to sensitivity analysis results from a traditional PVA.
4. Identify gopher tortoise life history parameters that may be amenable to manipulation via habitat management or that need additional research at Fort Stewart.

7.3   Model Description

NetLogo 4.0.4 (Wilensky 1999) was chosen for development of the IBM.[1] Vortex software (version 9.7) was used to generate the PVA and to help validate the IBM. NetLogo provides an accessible programming environment to support spatially explicit IBM simulation modeling (Wilensky 1999). Vortex is a population-based ecological modeling system (Lindenmayer et al. 1995). It captures dynamics of age-specific cohorts with respect to survival and reproduction statistics, and it also generates probabilities of metapopulation survival over time.

---

[1] An operational copy of this model is available through http://extras.springer.com.

Fig. 7.2 Map of Fort Stewart lands, depicting data layers used in individual-based model (IBM). *Grayscale shading* corresponds to estimated carrying capacity (with *darker shading* for higher carrying capacity and *white* for unsuitable patches). Gopher Tortoise Management Areas (GTMAs) are shown as *polygons* with *solid boundaries* and indicate the known concentrations of gopher tortoises. Army ranges are delineated with *double dashed lines*

7.3.1 Purpose

Our purpose was to assess the demographic factors most likely to influence persistence and abundance of gopher tortoises across Fort Stewart's range lands. However, the most influential (i.e., "sensitive") parameters were not necessarily the ones most amenable to manipulation, but some demographic traits were known to be influenced by habitat quality and thus could potentially serve as management targets. Ultimately, the purpose of the model was to guide management of gopher tortoises at the installation level by determining whether current habitat conditions and management practices would be likely to ensure the species' continued persistence on the installation, and if not, to evaluate whether increased habitat manipulation would be likely to influence demographic parameters sufficiently to improve overall species viability.

7.3.2 State Variables and Scales

7.3.2.1 Spatial and Temporal Scale

The spatial extent of the model's landscape is Fort Stewart (Fig. 7.2) which is the largest Army installation east of the Mississippi River (113,090 ha or approximately 54 × 30 km). Each patch in our model corresponded to a 150 × 150 m (2.25 ha) area. This size was large enough to allow us to model across the entire Fort Stewart landscape and for each patch to support more than one tortoise, yet not so large as to preclude movement among patches by individual tortoises during their lifetime. Each time step in the model represented 1 month to allow movement of individual tortoises between patches and to capture seasonal differences in movement probability (active vs. inactive season). Data were output from the model at 20-year intervals over the course of 100-year simulations.

7.3.2.2 Patch Variables

The primary patch variable was carrying capacity which is the number of adult tortoises each patch is predicted to support based on the soils and canopy cover within the patch (see Sect. 7.3.6). In this context, carrying capacity serves as a proxy for habitat quality. Gopher tortoises prefer deep sandy soils in which to burrow, and soil types have been previously classified as to their suitability for gopher tortoises (McDearman 1995; Hermann et al. 2002). In addition, burrow densities (and presumably tortoise densities) have been documented as varying with soil suitability (Jones and Dorr 2004). Canopy cover also has been shown to exert a strong influence on habitat selection by tortoises and thus is an important component of habitat carrying capacity (Aresco and Guyer 1999; Jones and Dorr 2004; Tuberville et al. 2007).

Patches also were characterized by whether a certain percentage of their area was comprised of the following features: (a) wetlands which serve as barriers to movement by tortoises, (b) active firing ranges and tank maneuver areas (hereafter referred to collectively as ranges) many of which are inaccessible and therefore have not been surveyed or do not have the potential for future gopher tortoise management due to this inaccessibility, and (c) Gopher Tortoise Management Areas (GTMAs) which correspond to delineated areas on Fort Stewart known to support concentrations of gopher tortoises (see Fig. 7.2 for ranges and GTMAs). The number of successful recruits (egg-to-juvenile stage) also was a patch variable which was based on the number of females that occupied each patch, the probability that a particular female reproduced in that year, and clutch size (which was affected by patch quality).

7.3.2.3 Individual (Agent) Variables

Agent-based variables were selected and parameterized to represent demographic traits associated with individual gopher tortoises. Individuals were classified as one of two types of agents (juveniles and adults) that behaved differently in the model. Mortality probability, dispersal probability, and dispersal distance varied between juveniles and adults. Juveniles graduated to adult status at age 15 years. Default parameter values were based on published values for gopher tortoises from Fort Stewart or the southeastern Georgia region when possible, or from data collected elsewhere in the species' range (Table 7.1). When published data were not available

Table 7.1 Individual parameters in the gopher tortoise model, their default values, and additional values tested to evaluate demographic sensitivity

Demographic parameter	Default	Other parameter values tested	References
Longevity	100	60, 70, 80, 90	Miller et al. (2001)
Age of sexual maturity (years)	15	None	
Mortality probability (annual)[a]			
Egg stage to age 1 (combined egg and hatchling mortality)	92	90, 91, 93, 94, 95, 96	Alford (1980), Landers et al. (1980), Pike and Seigel (2006), Tuberville et al. (2009)
Juvenile	13.5	9, 18, 22.5, 27	Modified from Tuberville et al. (2009)
Adult	1.5	3, 6, 9, 12	Ashton and Burke (2007), Tuberville et al. (2008), Guyer (unpublished data)
Dispersal probability (annual)[b]			
Juvenile	10	None	
Adult	2	None	Eubanks et al. (2003)
Reproduction			
Average clutch size[c]	6	4, 5, 7, 8	Rostal and Jones (2002), Mitchell (2005)
Proportion of females breeding (%)	80	60, 70, 90, 100	Rostal and Jones (2002), Mitchell (2005)
Sex ratio of clutch	0.5	None	

Tested values are based on the references provided

[a] Annual mortality rates were converted to monthly mortality rates in NetLogo, assuming mortality could occur with equal probability in any month of the year including when tortoises are relatively inactive

[b] Annual dispersal rates were converted to monthly dispersal rates, assuming dispersal occurred with equal probability in any month in which movement by tortoises is likely to occur (April–October in this model)

[c] For any given simulation, average clutch size also varies with habitat suitability, as specified in the reproduction submodel

for a particular parameter, default values were chosen based on values used in previous demographic models developed for gopher tortoises (Miller et al. 2001; Tuberville et al. 2009). Additional parameter values were tested to investigate model sensitivity by manipulating a single variable at a time and then comparing model results.

7.3.3   Process Overview and Scheduling

This model proceeded in monthly time steps for the duration of the 100-year simulations. Although 100 years may only represent at most three tortoise generations, our objectives relied on testing a timeframe more realistic to land and wildlife management goals. Based primarily on life history data available for the species in the literature, our model specified that the following processes (see also Sect. 7.3.7) occur each year during the month(s) indicated. An individual can disperse from or be evicted from its patch and attempt to relocate to another patch (April–October), assess the relative quality of its patch compared to surrounding patches (May) (see Sect. 7.3.7.2), reproduce (June), age (December), and die (January–December). In addition, mortality in the first year of life (egg stage-to-age 1) occurred in September.

7.3.4 Design Concepts

7.3.4.1 Emergence

The life cycle and behaviors of individuals were explicitly modeled through simple empirical rules governing the processes of mortality, reproduction, dispersal, and movement between patches. Population dynamics emerged from the collective behavior of individuals in the landscape. Emergent population dynamics included total population size, change in population size over the course of the simulation, and probability of extinction. In addition, population regulation emerged through the interaction between agents and patches; when the number of adult tortoises within a patch exceeded the patch's carrying capacity, the youngest adult tortoise was forced to move from the patch until the number of adult tortoises inside in patch was at or below carrying capacity.

7.3.4.2 Adaptation and Fitness

In the NetLogo model, each tortoise annually (in May) assessed the carrying capacity of its current patch relative to the eight neighboring patches. It then moved to (or remained in) the patch with the highest carrying capacity as long as the patch had space available (i.e., carrying capacity [number of adults in patch] > 0). Sufficient differences in carrying capacity (a proxy for habitat quality) between patches translated into differences in clutch size (Ashton et al. 2007). Finally, tortoises were allowed to move during any month of the activity season (April–October), but if they were not able to find a suitable patch with space available by the end of the activity season, they were forced to die in November. Thus, adaptation and fitness seeking were not explicitly modeled but resulted from the empirical rules governing individual behavior.

7.3.4.3 Sensing

Individual tortoises were able to sense the quality of their current and neighboring patches, and then behaved according to the movement, mortality, and reproductive rules specified for their individual sex and life stage in the model.

7.3.4.4   Interaction

The only interaction assumed to occur in the model was when carrying capacity was exceeded, the youngest adult was evicted from the patch and forced to search for another suitable patch with space available.

7.3.4.5   Stochasticity

The primary demographic and behavioral parameters including dispersal, mortality, and reproduction were interpreted as probabilistic processes. This approach was chosen because the default parameter values used in the model were based on population-level data from the literature and because we were interested in emergent population-level phenomena. Randomization was incorporated into the model during initialization of tortoises at the start of each simulation with sex, age, and location being assigned randomly according to the criteria described in Sect. 7.3.5. For each parameter combination (scenario), we ran 100 replicate simulations from which we calculated means from the response variables that were output from the model. Temporal environmental stochasticity (biotic or abiotic) was not incorporated into the model due to high levels of uncertainty that would confound assessments of demographic sensitivity.

7.3.4.6   Observation

We were interested in population-level variables such as the change in population size over course of the simulation and the probability of extinction. As part of model verification (sensu Rykiel 1996), during simulation runs the user interface plotted the tortoise population (juvenile, adult, total), tortoise density (overall and in suitable habitat only), and age distribution of the population. User-interface plots can be updated at each time step, or as with our models, updated annually in the time step corresponding to the month of reproduction (June) so that we could monitor population trends during simulations.

7.3.5   Initialization

The initial population size (3,000 tortoises) was based on an installation-wide survey for gopher tortoises conducted in 2009 (Macey, unpublished data), using the line transect, distance-sampling protocol described in Smith et al. (2009b). Because survey effort focused on those GTMAs that could be accessed by installation biologists, tortoises in the model were randomly placed among patches that were within GTMAs but outside of Fort Stewart's ranges. Although initial tortoise placement was restricted to outside the ranges, tortoises were allowed to subsequently move through ranges. Placement of tortoises in the landscape also was constrained by the model such that total number of tortoises populating a given patch was less than or equal to the carrying capacity of the patch. Finally, age and sex were randomly assigned to individuals such that the overall tortoise population had a 1:1 sex ratio and a normal adult age distribution with mean age of 30 years.

7.3.6   Input

The model inputs were raster and vector geographic information system (GIS) maps developed using the GRASS GIS (http://grass.itc.it). Raster maps were used to initialize patch variables and vector maps were used for visualization purposes. Raster maps included carrying capacity (generated from soils and canopy cover maps), wetlands, GTMAs, ranges, and study area boundaries. Vector maps included roads, streams, GTMAs, ranges, and study area boundaries. The primary model input was projected carrying capacity which was used as a proxy for habitat quality. To estimate the carrying capacity of each patch for gopher tortoises, we first generated a soils suitability map and a tree basal area map for Fort Stewart. The soils suitability map was created by reclassifying the Fort Stewart soils map previously digitized from 1:20,000 scale county soil survey maps published by the U.S. Department of Agriculture (USDA) Soil Conservation Service (SCS). Soils were reclassified as marginal, suitable, or priority soils based on established criteria (McDearman 1995; Guyer, Johnson, and Herman (unpublished data)).

Basal area was derived from the 2001 Gap Analysis Program (GAP) canopy cover map developed by the Multi-Resolution Land Characteristics (MRLC) Consortium (http://www.mrlc.gov/multizone_download.php?zone=14) using (7.1) reported for Michigan oak and pine stands (Buckley et al. 1999). We used average oak and pine canopy cover estimates from GAP maps for the Fort Stewart area to generate a basal area map for the installation, with resolution of 30 m and basal area output in units of square meters per hectare.

$$\text{Oak stand basal area} = [(\text{canopy cover} + 1.25) / 15.5] \qquad (7.1)$$
$$\text{Pine stand basal area} = [(\text{canopy cover} + 2.91) / 12.14]$$

In addition, a carrying capacity map with 30-m resolution was generated from the soils suitability map and the derived basal area map by using the formulas in (7.2), taken from Guyer, Johnson, and Herman (unpublished data).

$$\text{``Priority'' soils} = 9.7 \text{ tortoises/ha} * [100 - (1.43 * \text{basal area})] / 100$$
$$\text{``Suitable'' soils} = 2.9 \text{ tortoises/ha} * [100 - (1.43 * \text{basal area})] / 100 \qquad (7.2)$$
$$\text{``Marginal'' soils} = 1.2 \text{ tortoises/ha} * [100 - (1.43 * \text{basal area})] / 100$$

The 30-m resolution carrying capacity map was resampled at 150×150-m resolution to create the final projected carrying capacity used as input in the model. Vector maps of the installation boundary, GTMAs, ranges, and wetlands were converted to 30-m resolution raster maps and similarly resampled to create input maps with a 150×150-m resolution.

Throughout the model, we assumed carrying capacity related only to the number of adult tortoises in a patch. Juvenile tortoises and their burrows routinely are underestimated when using standard survey methods, due to their small size and cryptic appearance (Smith et al. 2009a; Tuberville and Dorcas 2001). In addition, habitat quality and social factors are both likely to influence the carrying capacity of a patch, with the latter factor presumably more likely to affect adults than juveniles. Therefore, the carrying capacity input map was used both in initializing tortoises in patches at the start of each simulation and in dictating eviction of the youngest adult tortoises from patches when carrying capacity was exceeded during the simulation.

Finally, a habitat class map was derived by categorizing individual patches in the carrying capacity map into the following habitat quality classes based on projected tortoise densities per hectare: habitats were classified as unsuitable (<0.5 tortoises/ha), low (0.5–2.9), moderate (3.0–5.0), and high quality (≥5). The habitat class map was used to apply reproductive penalties (reduced clutch sizes) to tortoises occupying lower-quality habitat patches (Ashton et al. 2007).

7.3.7 Submodels

7.3.7.1 Eviction from Patch

Adult tortoises were forced to move from their patch when carrying capacity within the patch was exceeded. The model evicted the youngest adult until the patch was again at or below carrying capacity. Once a tortoise was forced to leave a patch, the individual assessed the immediately surrounding eight patches and moved to the patch with the greatest space available (i.e., the greatest difference between carrying capacity and number of adult tortoises currently occupying the patch). If none of the neighboring cells had space available, the evicted tortoises moved to a randomly selected neighboring patch and continued to search for available space. The evicted tortoise could make up to eight attempts to find space available in neighboring patches per monthly time step, corresponding to a maximum cumulative movement distance of 1.2 km/month. The only additional constraint on movement in the eviction submodel was that tortoises, although allowed to move through ranges, could not move through patches that were classified as wetland. Eviction and the resulting search for patches with available space could occur in any month in which tortoise movement occurs (April–October).

7.3.7.2 Search for Better Habitat

Once per year, tortoises had the opportunity to relocate to better habitat. Every May, each individual assessed whether any of the neighboring patches had space available and then compared the habitat quality (i.e., carrying capacity) of those neighboring patches relative to its current patch. If habitat quality was highest in its current patch, the individual remained in the patch. If habitat quality was higher in one or more neighboring patches with space available, it would relocate to the neighboring patch with the highest habitat quality. Habitat quality (i.e., carrying capacity) was maintained as a static feature in our model based on the presumption that current habitat management efforts for gopher tortoises would continue and because tortoises in our model were ejected from a patch when carrying capacity was exceeded, thereby preventing resource depletion. Thus, as long as an individual did not move from its patch, an individual's associated habitat quality did not change during model simulations. However, other factors in our model elicited movement among patches by tortoises. This submodel provided tortoises the opportunity to respond to a heterogeneous landscape by moving among patches based on habitat quality and resource availability. In reality, gopher tortoises may elect to search for better habitat in any month during the active season. However, to significantly reduce simulation run time, we constrained the model so that this behavior was only allowed to occur in 1 month (May).

7.3.7.3 Dispersal

Gopher tortoises in high-quality habitat have small home ranges generally defined as 1–2 ha for adult males and <0.5 ha for adult females (Diemer 1992; Eubanks et al. 2003; Smith et al. 1997). While gopher tortoises occasionally will make long-distance excursions outside their home range, they will return to their core use areas. However, each year a small percentage of tortoises will disperse from their home range and establish a new home range in another location (Eubanks et al. 2003). We incorporated dispersal behavior into the model, assuming that dispersal could be motivated by factors other than habitat quality or carrying capacity of the current patch. In the dispersal submodel, a certain number of juvenile and adult tortoises were randomly selected to disperse based on previously defined dispersal probabilities. Dispersing tortoises oriented in a random direction and searched for patches with space available and occupied by at least one other tortoise. The maximum dispersal distance varied between adults and juveniles, but if individuals could not find an occupied patch within that distance, they were forced to stop. If tortoises encountered wetland patches while dispersing, they were forced to continue searching and move through patches that were not dominated by wetlands. Dispersal could occur during any month from April–October.

7.3.7.4 Reproduction

In the reproduction submodel, females produced eggs in June, which is the peak nesting month for gopher tortoises in southern Georgia (Landers et al. 1980). Adult females were randomly selected to reproduce according to the reproductive probability specified in the model. Clutch size varied among females, based on the habitat quality of their patch. Females in high and moderate quality patches produced the average clutch size specified in the model. Females in patches labeled as "low quality" incurred a reproductive penalty by producing two fewer eggs than the average clutch size. Females in unsuitable patches were not allowed to reproduce. Eggs hatched in September.

7.3.7.5 Aging

The age of individuals increased 1 year every December. After that time, individuals of the appropriate age graduated to the next life stage.

7.3.7.6 Mortality

A certain percentage of gopher tortoises died in each monthly time step based on stage-specific mortality probabilities for egg-to-age 1, juveniles, and adults. Monthly mortality probabilities were converted from the annual mortality probabilities specified in Table 7.1, assuming mortality could occur with equal probability in any month of the year including when tortoises were relatively inactive. In addition, any tortoises unable to locate suitable habitat patches by October were not allowed to overwinter there and were forced to die in November. Finally, tortoises were not allowed to live more than 100 years in the model.

7.4 Simulation Experiments

7.4.1 Model Calibration

As part of the model calibration process (sensu Rykiel 1996), we adjusted the default value for juvenile survivorship (the parameter that is least well known) until the combination of default parameters resulted in a sustained population. Initially, we set juvenile mortality to 15% based on previous work by Tuberville et al. (2009). However, we found that a 10% reduction was required to produce a stable population in NetLogo (i.e., for a juvenile, a resulting mortality rate of 13.5% was set as the default value). Once we selected the combination of default values for our IBM, we varied parameters individually over a range of biologically realistic values (Table 7.1) to perform demographic sensitivity analysis.

7.4.2 Model Validation

We intended to validate the demographic sensitivity results of our IBM by performing a similar analysis using the more traditional PVA approach. We constructed a

PVA in Vortex 9.7 by using the default and range-of-parameter values identified in Table 7.1 and systematically varying each parameter, one at a time. Each scenario (i.e., parameter combination) was run for 100 years and repeated for a total of 100 simulations. Each simulation was initialized with a starting population of 300 tortoises, characterized by an age structure that mirrored the starting age structure in the IBM simulations. We selected a starting population size of 300 tortoises (10% of the population estimate for Fort Stewart that was used as the initial population size for the IBM simulations), which was based on minimum reserve sizes estimated by McCoy and Mushinsky (2007) and Styrsky et al. (2010), and the number of tortoises predicted to occur on those reserves. Although Vortex will support models with multiple populations linked in a metapopulation structure, we did not have sufficient data for Fort Stewart to estimate dispersal among GTMAs. Thus, our Vortex simulations were not spatially explicit and assumed a single cohesive population rather than several "populations" distributed among habitat patches across the landscape. For each PVA scenario, we reported the average population size at 20-year intervals and the probability of extinction over the 100-year simulation. To scale our demographic sensitivity results from the PVA to results from the IBM, we also reported the percentage change in population size for both sets of results.

7.4.3   Effects on Population Trends and Probability of Extinction

Simulations of the baseline scenario (i.e., all parameters set to default values) for our IBM resulted in an average population increase of 7.1% (Table 7.2). Any scenario in which a single parameter was set at a value less favorable than in the baseline scenario predicted a population decline during the 100-year simulations, although rarely did it result in extinction (defined here as when fewer than two individuals remain at end of simulation). In fact for our IBM model, the probability of extinction (P_E) was greater than zero only when annual adult mortality was set to 9 or 12%, which resulted in a P_E of 0.11 and 0.87, respectively.

Compared to IBM simulations, PVA simulations of the same scenario were less likely to predict a population decline than IBM simulations (Table 7.2). However, for those scenarios where the PVA *did* predict a decline, the P_E tended to be higher for PVA than for IBM. When both IBM and PVA simulations projected a positive change in population size (e.g., when longevity was set to 90 years or juvenile mortality to 9%), the PVA predicted a greater percentage increase than did the IBM; the converse was observed when both the model types projected a negative change in population size (e.g., when adult mortality was set to 3% or proportion of females breeding to 60%). When the direction of population change differed between the two model types (i.e., positive or negative), the IBM predicted a population decrease while the PVA predicted a population increase. Finally, the PVA model exhibited a greater magnitude of responses among scenarios than did the IBM.

Table 7.2 Comparison of IBM and PVA model results based on same set of scenarios and parameter combinations

Scenario name	IBM results (N_0 = 3000 tortoises)			PVA results (N_0 = 300 tortoises)		
	Final N	Change (%)	P_E	Final N	Change (%)	P_E
Baseline (all default parameters)	3,215	+7.2	0	544	+81.4	0
Longevity = 90 years	3,092	+3.1	0	482	+60.6	0
Longevity = 80 years	2,891	−3.6	0	411	+37.1	0
Longevity = 70 years	2,674	−10.9	0	311	+3.5	0
Longevity = 60 years	2,357	−21.4	0	207	−30.9	0
Adult mortality = 3%	899	−70.0	0	189	−36.9	0
Adult mortality = 6%	78	−97.4	0	26	−91.3	0.02
Adult mortality = 9%	7	−99.8	0.11	8	−97.3	0.62
Adult mortality = 12%	<1	−100.0	0.87	<1	−100.0	1.00
Juvenile mortality = 9%	10,213	+240.4	0	2,721	+806.9	0
Juvenile mortality = 18%	1,497	−50.1	0	134	−55.3	0
Juvenile mortality = 22.5%	925	−69.2	0	40	−86.6	0
Juvenile mortality = 27%	701	−76.6	0	14	−95.5	0.03
Egg-to-juvenile mortality = 90%	4,907	+63.6	0	912	+204.1	0
Egg-to-juvenile mortality = 91%	4,003	+33.4	0	705	+135.0	0
Egg-to-juvenile mortality = 93%	2,554	−14.9	0	409	+36.4	0
Egg-to-juvenile mortality = 94%	2,004	−33.2	0	279	−7.0	0
Egg-to-juvenile mortality = 95%	1,551	−48.3	0	198	−33.9	0
Egg-to-juvenile mortality = 96%	1,178	−60.7	0	133	−55.8	0
Mean clutch size = 8	5,844	+94.8	0	1,070	+256.5	0
Mean clutch size = 7	4,374	+45.8	0	796	+165.3	0
Mean clutch size = 5	2,321	−22.6	0	362	+20.9	0
Mean clutch size = 4	1,645	−45.2	0	225	−25.0	0
Proportion females breeding = 100%	4,690	+56.3	0	917	+205.8	0
Proportion females breeding = 90%	3,881	+29.4	0	704	+134.6	0
Proportion females breeding = 70%	2,625	−12.5	0	398	+32.7	0
Proportion females breeding = 60%	2,133	−28.9	0	286	−4.7	0

Scenarios are grouped according to parameter manipulations and are ordered from most favorable to least favorable conditions within each parameter. All IBM simulations start with initial population size of 3,000 tortoises; PVA simulations start with 300 tortoises. For each model type, the following are reported: gopher tortoise population size at end of simulation (Final N), percent change in population size during simulation (% change), and probability of extinction (P_E; range of possible values 0–1). *Shaded cells* correspond to simulations resulting in population decline and/or $P_E > 0$

7.4.4 Demographic Sensitivity Analysis

For each model type, we evaluated sensitivity to changes in six demographic parameters by individually varying those parameters over a range of biologically realistic values. However, due to expected differences in plasticity of corresponding life history traits, the range of values tested varied among parameters. In order to relate the magnitude of differences in parameter values tested to the magnitude of resulting responses, we calculated the following two ratios.

Parameter ratio = scenario parameter value/baseline parameter value
Response ratio = scenario population size/baseline population size

For example, for the scenario where adult mortality = 3%, the parameter value for the scenario was 3% compared to the baseline scenario's default value of 1.5%, resulting in a parameter ratio of 2. Based on the IBM, the predicted final tortoise population for the scenario of interest was 899, compared to a final population size of 3,215 in the baseline scenario (Table 7.2). This scenario resulted in a response ratio of 899/3,215 or 0.28. Based on these ratios, the IBM predicted that a doubling of the baseline adult mortality rate would result in a 78% smaller ending population when compared to a scenario using default values. For the baseline scenario, the parameter ratio and the response ratio always were equal to 1.

Results of these ratio calculations are shown in Fig. 7.3, where the parameter ratio (independent variable) was plotted on the x axis, and the response ratio (dependent variable) was plotted on the y axis; the steepness of the curve indicates the sensitivity of the model to changes in parameter value. Ratios for each of the six demographic parameters were plotted separately, with ratios for both the IBM and PVA model presented in the same graph. All results were plotted on the same scale to facilitate comparisons among parameters.

Despite differences in population size and percentage change in population size observed between our IBM and PVA model (Table 7.2), the two model types exhibited remarkable congruence in their sensitivities to manipulation of demographic variables. In fact, the results are so similar that it sometimes is difficult to discern that two distinct data sets are plotted (Fig. 7.3). The one notable exception is juvenile survivorship, which predicted similar responses in the IBM and PVA at parameter values greater than or equal to baseline value; although both model types predicted dramatically larger populations when juvenile mortality rates were lower than baseline, the PVA model predicted a greater response than the IBM.

As mentioned previously, we varied values for each demographic parameter over a range of biologically realistic values. Keeping that in mind, several observations can be made from our results.

1. The baseline values for adult mortality and longevity represented the most optimistic scenario (i.e., no scenario has a parameter ratio greater than 1); all other parameter values predicted smaller population sizes (i.e., response ratio less than 1).
2. Scenarios varying juvenile mortality exhibited the greatest magnitude in responses among scenarios over the range of values tested.

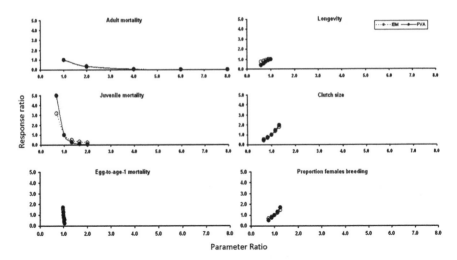

Fig. 7.3 Comparison of demographic sensitivity analysis results between IBM (*dashed lines, open circles*) and population viability model (*solid lines* and *circles*). For each demographic parameter, the parameter ratio (scenario parameter value/baseline parameter value) is plotted on the *x*-axis and the response ratio (scenario population size/baseline population size) on the *y*-axis. Note that results for all parameters are plotted at the same scale

3. Based on the nearly vertical plots for egg-to-age 1 mortality, this demographic parameter predicted the greatest difference in population size per unit difference (i.e., the models were most sensitive to changes in this parameter).

7.5 Discussion

7.5.1 *Population Trends and Probability of Extinction*

Although we reported final population sizes and probability of extinction after 100 years for each scenario tested, the primary goal of the IBM and PVA models was to identify parameters to which our simulated populations were most sensitive rather than to predict the specific outcome of any given scenario. Therefore, one must be cautious about interpreting and applying the results of our models to specific management scenarios for gopher tortoises on Fort Stewart. However, our results are useful for identifying patterns and potential factors contributing to those patterns.

One major pattern we observed was that when any demographic parameter (except longevity) was set to a value less optimistic than the baseline scenario, the IBM predicted a population decline. In the case of adult mortality, the default value was the most optimistic value, suggesting that any chronic increase in adult mortality above baseline would result in declines that could potentially compromise long-term persistence if all other parameters remained the same. The importance of high adult survivorship in the life history of most turtle species and in the management of individual populations is well documented (e.g., Congdon et al. 1993, 1994). We did not run simulations with adult mortality less than 1.5% because we wanted to only test parameters with biologically realistic values. The most optimistic value used in our simulations was based on studies of gopher tortoise populations on protected lands (Ashton and Burke 2007; Tuberville et al. 2008), and we are aware of no published study to date that has documented lower mortality rates.

The other major pattern that emerged was that the PVA was more likely than the IBM to predict a population increase and to predict an increase of greater magnitude. Apparent discrepancies between the two model types are probably best explained by additional challenges imposed by the spatial context of the IBM modeling environment. The PVA we constructed in Vortex assumed a single population that was closed to immigration and emigration, and in which all individuals were exposed to the same habitat conditions. Furthermore, carrying capacity for the entire population was set at ten times the initial population size (i.e., 3,000).

In contrast, carrying capacity was specified for each patch (i.e., pixel) in the IBM. Carrying capacity could have affected overall population in at least two ways. First, the sum of carrying capacity values for individual patches determined the maximum population size possible for the entire installation (at least for adults), although population sizes in our simulations never approached this maximum value. More importantly, however, carrying capacity of individual patches influenced the initial placement of tortoises and also determined whether tortoises were forced to leave a patch to search for suitable habitat elsewhere. NetLogo initiated each simulation by placing tortoises in suitable patches among GTMAs, which were distributed across Fort Stewart and interspersed with unsuitable habitat, effectively fragmenting the installation's tortoise population. In addition, when carrying capacity for an individual patch was exceeded, individual tortoises were ejected from the patch and forced to search for a suitable patch that had not yet reached carrying capacity. In some cases, individuals were not able to find a patch meeting these criteria before activity season ended, at which time they were forced to die. Thus, the landscape context of the IBM (particularly when that landscape is patchy) created a more complex real-world environment for the simulated population. It is likely the landscape context also contributed to the differences in population trends that we observed between the two types of models.

7.5.2 Demographic Sensitivity Analysis

When considering sensitivity to individual parameters, the congruence between model types suggests that IBMs, such as the one we developed for gopher tortoises in NetLogo, can be a valuable approach to conducting demographic sensitivity analysis. Because the overall sensitivity results from the IBM and PVA models were so similar, we will restrict our discussion to the IBM.

Although scenarios in which adult mortality rates were elevated above baseline level were the only scenarios likely to result in population extinction (Table 7.2), adult mortality was not the parameter to which the model was most sensitive (Fig. 7.3). In addition, adult mortality rate in the baseline scenario (1.5%) was the most optimistic value tested. The baseline value corresponded to long-term estimates that have been published about two different translocated populations (Ashton and Burke 2007; Tuberville et al. 2008). We could find no estimates of long-term adult mortality rates for any naturally occurring gopher tortoise populations in the peer-reviewed literature, but the estimate used in our model is among the lowest adult mortality estimates reported for chelonians (Iverson 1991; Wilbur and Morin 1988). From an application perspective, these findings imply that although it may not be feasible to reduce adult mortality below 1.5%, monitoring and management efforts should prevent or mitigate threats that could potentially cause chronic increases in adult mortality.

The parameter associated with the greatest magnitude of change in population size over the values tested was juvenile mortality. The lowest juvenile mortality rate (9%) predicted population sizes that were three times the size predicted by the baseline scenario in which juvenile mortality was set to 13.5%. Juvenile mortality rates could conceivably be reduced through habitat management particularly in areas where canopy cover is excessively high or herbaceous vegetation is limited. Low canopy cover (≤60%) and basal area (30 m²/ha) are important for providing necessary thermal conditions for tortoises and for promoting diverse and abundant herbaceous vegetation in the understory (Aresco and Guyer 1999; Tuberville et al. 2007; Wilson et al. 1997). In turn, the herbaceous understory provides forage for growth and camouflaging cover from predators while vulnerable juveniles are active outside their burrows. We are aware of no studies comparing juvenile survivorship rates among different habitat types; indeed, few data at all are available for estimating juvenile survivorship (but see Wilson 1991; Tuberville et al. 2008; and inferred estimates from Pike et al. 2008). Obtaining information on juvenile tortoise mortality rates from Fort Stewart would greatly improve the predictive value of the IBM and its utility for guiding management of gopher tortoise populations on the installation.

The parameter for which the smallest change predicted the greatest effect on population size was egg-to-age 1 mortality. Mortality rates for both eggs and hatchlings are known to be high and are suspected—at least for the egg stage—to be quite variable among sites and years (Landers et al. 1980; Pike and Seigel 2006; Wright 1982). Such high and variable mortality rates highlight the importance of this parameter in the model and the challenge of making broadly applicable management recommendations. Mortality in the egg-to-age 1 stage probably can be influenced by habitat management in the same manner that we suspect juvenile mortality may be influenced. Egg-to-age 1 mortality also probably is influenced to a greater extent by environmental stochasticity and predator population cycles than is juvenile mortality. Thus, although habitat management may improve conditions such that hatchling survivorship increases, other factors more difficult to control may hinder the ability of habitat management to have an appreciable effect on population level.

Finally, our IBM appears sufficiently sensitive to changes in both clutch size and proportion of females breeding to suggest that those two demographic traits merit consideration as management targets. For both parameters, we observed a twofold difference in population size between scenarios with the baseline vs. the maximum parameter values. Data from Fort Stewart and a nearby state park suggest that these demographic traits can vary among sites with differing habitat quality (Rostal and Jones 2002). It is therefore feasible that these demographic traits could be improved through habitat-based management approaches.

7.6 Conclusions and Recommendations

We found IBM to be a useful tool for performing demographic sensitivity analysis, and we observed remarkable agreement between sensitivity results from our IBM and PVA models for gopher tortoises on Fort Stewart. Our analyses identified several demographic traits that appeared to disproportionately influence size of simulated populations, particularly mortality rates in the juvenile stage and egg-to-age 1 stage. Although there are few data comparing these rates among sites with varying habitat quality, we suspect both traits are fairly responsive to changes in habitat quality and thus, habitat manipulation. In addition, it is important to keep in mind our analyses varied by only a single parameter at a time, whereas habitat improvement at poorer quality sites should positively influence a suite of demographic traits simultaneously. Furthermore, our baseline model, which was intended to reflect demographic traits for tortoises under current management conditions at Fort Stewart, predicted a population increase during the 100-year simulations. Based on these factors, we believe current management conditions are conducive to long-term persistence of gopher tortoises in the Fort Stewart landscape, and that habitat management is a practical and effective means to improve population conditions at poorer quality sites.

Should there be a net loss of suitable habitat for gopher tortoises—perhaps as a result of range construction or infrastructure development—resource managers likely will need to improve habitat conditions for tortoises on remaining patches. Most of Fort Stewart is comprised of soils considered to be unsuitable for gopher tortoises (Fig. 7.4a), which limits the ability of resource managers to implement large-scale improvements in habitat quality across the installation. However, approximately 13% of the landscape can be characterized as marginal habitat occurring on suitable soils (Fig. 7.4b). Habitat improvements targeting these patches could increase carrying capacity of individual patches, hopefully offsetting any potential losses in suitable habitat.

There are two important caveats to our conclusions that stem from the following assumptions in our model: (1) parameter values in our IBM are representative of demographic traits for gopher tortoises on Fort Stewart, and (2) current wildlife and land management practices will be maintained so that there is no change in the amount, quality, or distribution of suitable habitat patches across the installation.

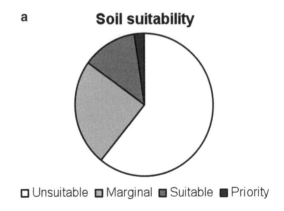

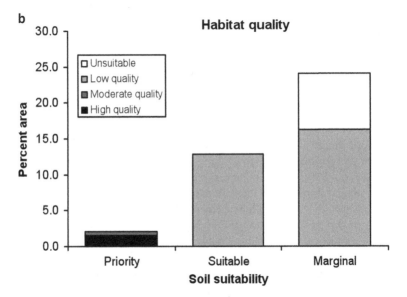

Fig. 7.4 Habitat conditions for gopher tortoises in the Fort Stewart landscape in terms of: (**a**) relative abundance of unsuitable, marginal, suitable, and priority soil types (categories based on information presented in McDearman 1995; Hermann et al. 2002), and (**b**) habitat quality of patches associated with each soil suitability category

We estimated parameters for our IBM from the literature, using data specific to Fort Stewart or the surrounding region whenever possible. However, for some parameters—particularly juvenile mortality and longevity—few data or no data were available. In addition, parameters estimated from data from other portions of the species' range may not reflect demographic traits for gopher tortoises on Fort Stewart. However, as more data become available, parameter estimates can be modified to refine the model and make it more useful to resource managers making installation-specific management recommendations. Resource managers also can adapt the model to examine gopher tortoise dynamics under changing or altered landscapes, such as conversion of tortoise habitat to training areas or habitat improvement in suitable patches outside of GTMAs. These landscape changes can affect not only the suitability of individual patches for gopher tortoises but also the distribution of suitable patches and the ability of tortoises to move between patches. By considering both demographic traits and behavior of individuals—particularly regarding how they interact with the landscape—IBMs offer a powerful means for resource managers to evaluate the likelihood of long-term persistence of rare species under current or alternative landscape conditions and to identify and implement appropriate management actions.

Acknowledgments This research was funded by U.S. Army Engineer Research and Development Center. Manuscript preparation was partially supported by the Department of Energy under Award Number DE-FC09-07SR22506. We thank the following for their assistance: Ron Owens at Fort Stewart provided GIS data layers for the installation, Jen Burton (ERDC) provided programming advice, and Bess Harris (SREL) assisted with running models.

References

Alford RA (1980) Population structure of *Gopherus polyphemus* in northern Florida. J Herpetol 14(2):177–182

Aresco MJ, Guyer C (1999) Burrow abandonment by gopher tortoises in slash pine plantations of the Conecuh National Forest. J Wildl Manage 63(1):26–35

Ashton KG, Burke RL (2007) Long-term retention of a relocated population of gopher tortoises. J Wildl Manage 71(3):783–787

Ashton KG, Burke RL, Layne JN (2007) Geographic variation in body and clutch size of gopher tortoises. Copeia 2:355–363

Buckley D, Isebrands JG, Sharik TL (1999) Practical field methods of estimating canopy cover, PAR, and LAI in Michigan oak and pine stands. J Appl Forest 16(1):25–32

Congdon JD, Dunham AE, Loben Sels RC (1993) Delayed sexual maturity and demographics of Blanding's turtles (*Emydoidea blandingii*): implications for conservation and management of long-lived organisms. Conserv Biol 7(4):826–833

Congdon JD, Dunham AE, Loben Sels RC (1994) Demographics of common snapping turtles (*Chelydra serpentina*): implications for conservation and management of long-lived organisms. Am Zool 34(3):397–408

Diemer JE (1992) Home range and movements of the tortoise (*Gopherus polyphemus*) in northern Florida. J Herpetol 26(2):158–165

Eubanks JO, Michener WK, Guyer C (2003) Patterns of movement and burrow use in a population of gopher tortoises (*Gopherus polyphemus*). Herpetologica 59(3):311–321

Hermann SM, Guyer C, Waddle JH, Nelms MG (2002) Sampling on private property to evaluate population status and effects of land use practices on the gopher tortoise, *Gopherus polyphemus*. Biol Conserv 108(3):289–298

Iverson JB (1991) Patterns of survivorship in turtles (order Testudines). Can J Zool 69(2):385–391

Jones JC, Dorr B (2004) Habitat associations of gopher tortoise burrows in industrial timberlands. Wildl Soc Bull 32(2):456–464

Landers JL, Garner JA, McRae WA (1980) Reproduction of gopher tortoises (*Gopherus polyphemus*) in southwestern Georgia. Herpetologica 36(4):353–361

Lindenmayer DB, Burgman MA, Akçakaya HR, Lacy RC, Possingham HP (1995) A review of the generic computer programs ALEX, RAMAS/space and VORTEX for modelling the viability of wildlife metapopulations. Ecol Model 82(2):161–174

McCoy ED, Mushinsky HR (2007) Estimates of minimum patch size depend on the method of estimation and the condition of the habitat. Ecology 88(6):1401–1407

McDearman W (1995) Gopher tortoise (*Gopherus polyphemus*) soil classification for the federally listed range. U.S. Fish and Wildlife Service, Mississippi Field Office, Jackson

Miller PS et al (2001) Preliminary population viability assessment for the gopher tortoise (*Gopherus polyphemus*) in Florida. In: Prepared by participants from gopher tortoise population viability workshop, Tallahassee, 11–12 Sept 2001 and Conservation breeding specialist group (Species Survival Commission/IUCN World Conservation Union), Apple Valley. http://www.cbsg.org/cbsg/workshopreports/23/gopher_tortoise_pva.pdf. Accessed September 2010

Mitchell M (2005) Home range, reproduction, and habitat characteristics of the female gopher tortoise (*Gopherus polyphemus*) in southeast Georgia. M.S. thesis, Georgia Southern University, Statesboro. http://eaglescholar.georgiasouthern.edu:8080/jspui/bitstream/10518/1588/3/Mitchell_Maggie_J_200508_MS.pdf. Accessed April 2010

NatureServe (2004) Species at risk on Department of Defense installations. Revised report documentation of study for Department of Defense and U.S. Fish and Wildlife Service. http://www.natureserve.org/prodServices/speciesatRiskdod.jsp. Accessed April 2010

Pike DA, Seigel RA (2006) Variation in hatchling tortoise survivorship at three geographic localities. Herpetologica 62(2):125–131

Pike DA, Pizzatto L, Pike DA, Shine R (2008) Estimating survival rates of uncatchable animals: the myth of high juvenile mortality in reptiles. Ecology 89(3):607–611

Reed JM, Fefferman N, Averill-Murray RC (2009) Vital rate sensitivity analysis as a tool for assessing management actions for the desert tortoise. Biol Conserv 142(11):2710–2717

Rostal DC, Jones DN Jr (2002) Population biology of the gopher tortoise (*Gopherus polyphemus*) in southeast Georgia. Chelonian Conserv Biol 4(2):479–487

Rykiel EJ Jr (1996) Testing ecological models: the meaning of validation. Ecol Model 90(3):229–244

Smith RB, Breininger DR, Larson VL (1997) Home range characteristics of radiotagged gopher tortoises on Kennedy Space Center, Florida. Chelonian Conserv Biol 2(3):358–362

Smith LL, Tuberville TD, Seigel RA (2006) Workshop on the ecology, status, and management of the gopher tortoise (*Gopherus polyphemus*), Joseph W. Jones Ecological Research Center, 16–17 January 2003: final results and recommendations. Chelonian Conserv Biol 5:326–330

Smith LL, Linehan JM, Stober JM, Elliott MJ, Jensen JB (2009a) An evaluation of distance sampling for large-scale gopher tortoise surveys in Georgia, USA. Appl Herpetol 6(4):355–368

Smith LL, Stober JM, Balbach HE, Meyer WD (2009b) Gopher tortoise survey handbook. Technical Report, ERDC-CERL, Champaign. Report No. ERDC/CERL TR-09-7. http://libweb.wes.army.mil/uhtbin/hyperion/CERL-TR-09-7.pdf. Accessed October 2010

Styrsky JN, Guyer C, Balbach H, Turkmen A (2010) The relationship between burrow abundance and area as a predictor of gopher tortoise population size. Herpetologica 66:403–410

Tuberville TD, Dorcas ME (2001) Winter survey of a gopher tortoise population in South Carolina. Chelonian Conserv Biol 4(1):182–186

Tuberville TD, Buhlmann KA, Balbach HE, Bennett SH, Nestor JP, Gibbons JW, Sharitz RR (2007) Habitat selection by the gopher tortoise (*Gopherus polyphemus*). ERDC/CERL TR-07-1. U.S. Army Engineer Research and Development Center, Champaign. http://libweb.wes.army.mil/uhtbin/hyperion/CERL-TR-07-1.pdf. Accessed October 2010

Tuberville TD, Norton TM, Todd BD, Spratt JS (2008) Long-term apparent survival of translocated gopher tortoises: a comparison of newly released and previously established animals. Biol Conserv 141(11):2690–2697

Tuberville TD, Gibbons JW, Balbach HE (2009) Estimating viability of gopher tortoise populations. ERDC/CERL TR-09-2. U.S. Army Engineer Research and Development Center, Champaign. http://libweb.wes.army.mil/uhtbin/hyperion/CERL-TR-09-2.pdf. Accessed October 2010

U.S. Fish and Wildlife Service (Department of the Interior) (1987) Endangered and threatened wildlife and plants; determination of threatened status for the gopher tortoise (*Gopherus polyphemus*). 50 CFR, Part 17, Final Rule, Federal Register 52(129):25376–25380

U.S. Fish and Wildlife Service (Department of the Interior) (2009) Endangered and threatened wildlife and plants; 90-day finding on a petition to list the eastern population of the gopher tortoise (*Gopherus polyphemus*) as threatened. 50 CFR, Part 17, FWS-R4-ES-2009-0029; MO 9221050083-B2, Federal Register 74(173):46401–46406

Wilbur HM, Morin PJ (1988) Life history evolution in turtles. In: Gans C, Huey R (eds) Biology of the reptilia, vol 16b. Alan R. Liss, New York, pp 396–447

Wilensky U (1999) NetLogo: computer software. Center for Connected Learning and Computer-Based Modeling, Northwestern University, Evanston. http://ccl.northwestern.edu/netlogo/. Accessed June 2009

Wilson DS (1991) Estimates of survival for juvenile gopher tortoises, *Gopherus polyphemus*. J Herpetol 25(3):376–379

Wilson DS, Mushinsky HR, Fischer RA (1997) Species profile: gopher tortoise (*Gopherus polyphemus*) on military installations in the southeastern United States. Technical Report SERDP-97-10. Waterways Experiment Station, U.S. Army Corps of Engineers, Vicksburg. http://el.erdc.usace.army.mil/tes/pdfs/serdp97-10.pdf. Accessed October 2010

Wright JS (1982) Distribution and population biology of the gopher tortoise, *Gopherus polyphemus*, in South Carolina. Thesis, Clemson University, Clemson

Chapter 8
A Model for Evaluating Hunting and Contraception as Feral Hog Population Control Methods

Jennifer L. Burton, Marina Drigo, Ying Li, Ariane Peralta, Johanna Salzer, Kranthi Varala, Bruce Hannon, and James D. Westervelt

8.1 Background

This model was developed to explore the relative effectiveness of controlling feral swine (*Sus scrofa*) with hunting, contraception, and a combination of hunting and contraception. Feral swine are an invasive species known to feed on small animals, eggs, roots, and herbaceous material. They alter the environment, disrupting plant

J.L. Burton (✉)
Department of Natural Resources and Environmental Sciences, University of Illinois,
W-503 Turner Hall, 1102 South Goodwin Ave, Urbana, IL 61801, USA
e-mail: jlburton@illinois.edu

M. Drigo
Department of Urban and Regional Planning, University of Illinois,
111 Temple Buell Hall, 611 Taft Drive, Champaign, IL 61820, USA

Y. Li • K. Varala
Department of Crop Sciences, University of Illinois, AW-101 Turner Hall,
1102 South Goodwin Avenue, Urbana, IL 61801, USA

A. Peralta
Program in Ecology, Evolution, and Conservation Biology, University of Illinois,
286 Morrill Hall, 505 South Goodwin Ave, Urbana, IL 61801, USA

J. Salzer
Department of Pathobiology, University of Illinois, 2522 VMBSB,
2001 S. Lincoln, Urbana, IL 61801, USA

B. Hannon
Department of Geography, University of Illinois, 220 Davenport Hall,
607 S Mathews, M/C 150, Urbana, IL 61801, USA

J.D. Westervelt
Construction Engineering Research Laboratory, US Army Engineer Research
and Development Center, Champaign, IL, USA
e-mail: james.d.westervelt@usace.army.mil

J.D. Westervelt and G.L. Cohen (eds.), *Ecologist-Developed Spatially Explicit*
Dynamic Landscape Models, Modeling Dynamic Systems,
DOI 10.1007/978-1-4614-1257-1_8, © Springer Science+Business Media, LLC 2012

and animal habitat, by rooting in the soil (Graves 1984). Feral swine are highly adaptable. As true dietary generalists, they can thrive on an extensive spectrum of food sources. They are also unusually prolific for large mammals (Mauget et al. 1991). A single female may be capable of producing more than 26 offspring in a single year. Adult feral swine have few predators, none of which is significant other than humans. Even after their numbers have been reduced by drought, food shortage, or lethal removal, their unique combination of adaptability and fecundity enables the population to recover quickly soon after the threat is removed (Dziecolowski et al. 1992). The animals' intelligence further complicates population control efforts: feral swine quickly learn to avoid areas where hunting or trapping have taken place (Stephen S. Ditchkoff, Auburn University, personal communication 2008).

Fort Benning, GA, is home to several at-risk species of fauna and flora, including the federally listed gopher tortoise (*Gopherus polyphemus*) and relict trillium (*Trillium reliquum*), both of which can be negatively impacted by feral swine predation and rooting activities. It is currently estimated that about 5,000 feral swine reside at Fort Benning, and this population is supplemented with immigration from other regional populations. Roughly 30,000 acres of the impacted area are off limits to humans because of hazards posed by unexploded ordnance. An ongoing bounty program led to the apparent removal of over 1,200 feral pigs from Fort Benning from June 2007 through February 2008. The program is believed to have some success in reducing the swine population despite their tremendous reproductive capacity. In research conducted at Fort Benning, Hanson (2006) concluded that both immigration and increased reproductive rates (in response to intensive removal activities) factor into the ability of this population to withstand substantial lethal removal without a decrease in population growth rates. Therefore, the bounty program alone may not be sufficient to remove feral hog threats to remaining gopher tortoise and relict trillium populations.

In addition to the hunting program operated by Fort Benning for lethal removal of feral swine, other proposed options include trapping, poisoning, physical exclusion, and contraceptive delivery. The feasibility and effectiveness of each control option is influenced by cost, mode of delivery, route of administration, specificity for the target population (e.g., species, sex), and—for nonlethal controls—duration of effect. Physical exclusion, such as fencing, is considered impractical due to the length and nature of the terrain at the Fort Benning perimeter.

8.2   Objective

The objective of this project was to answer the following questions about feral pig demographics, terrain, and hunting data specific to Fort Benning:

- How would a contraceptive program affect the density of the Fort Benning feral swine population over time?
- Could a contraceptive program replace lethal control for this population?
- How would the combination of sterilization and lethal control programs affect the density of the feral pig population?

We hypothesized that swine population control would be optimized by a combination of lethal removal and contraceptive delivery.

8.3 Model Description

8.3.1 Purpose

In order to evaluate the potential for control of a feral pig population via hunting and/or contraception, we constructed a spatially explicit model of feral pig demographic and spatial characteristics relative to specific features of the Fort Benning landscape using the agent-based modeling system NetLogo 4.0.2 (Wilensky 1999).[1] Our model was populated with the best available information about *Sus scrofa* in general and the Fort Benning population in particular, but it is designed to be readily adaptable to different feral swine populations, control methods, and sites. The model covers an area that includes all of Fort Benning and Columbus, GA, and captures processes that define landscape characteristics (land cover, land use, vegetation, and water availability), individual hogs, hunting, and contraceptive baits (Fig. 8.1). Dynamic hog behavior captures reproductive cycles, attrition, social grouping and dynamics, diet and feeding, and movement. Hog hunting and distribution of contraceptive baits were modeled as individual and combined options for hog control. The equations, parameters, and variables that define the model are taken from the literature with expert advice from Fort Benning environmental staff familiar with the feral hog populations there. The model is described below.

8.3.2 State Variables and Scales

The landscape for this model consists of a grid of 206×213 cells, with each cell representing a 200×200-m patch. The model operates on a 1-week time step. The primary agents in this model are landscape patches and pigs. A third agent, called a *sounder*, represents a group of female pigs; it facilitates moving pigs together around the landscape. Each sounder agent has several state variables: a list of associated pigs, the week that that group *farrows* (i.e., gives birth), and travel in the current time step. The pig agent also manages several state variables: age, sounder, travel in the current time step, and color. The color is used to identify the individual as male (blue), female (red), or sterilized female (yellow). Each patch is defined by state variables that contain the following information:

- Whether the cell is accessible to pigs
- Whether the cell is accessible to humans

---

[1] An operational copy of this model is available through http://extras.springer.com.

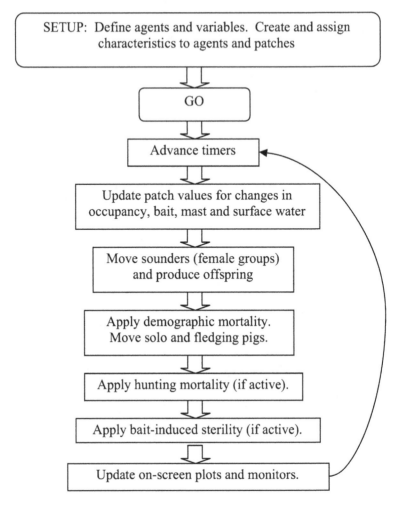

Fig. 8.1 Model overview

- Whether contraceptive bait exists in the patch
- Distance to water in the dry season
- Distance to water in the wet season
- Potential for generating mast (i.e., hard-shelled seeds and nuts)
- Current availability of mast
- Optimal attractiveness of the patch to pigs
- Current attractiveness based on optimal and presence of others
- Sounder whose sign is strongest in the patch
- Boar whose sign is strongest in the patch
- Representation of other pig sign strength

8.3.3   Process Overview and Scheduling

The model is initialized by importing patch variables from digital maps of Fort Benning and distributing an initial population of feral hogs on the landscape. At each time step (i.e., 1 week), the following events occur:

1. The state of the patches is set based on the time of year, time remaining on bait viability, and current presence of boars and sounders.
2. The sounders are updated, which may involve deleting those no longer containing members, allowing them to move, and farrowing.
3. All pigs are allowed to potentially die, age, move, fledge, and regroup into sounders.

The simulation time is divided into pre-control and control periods. The pre-control period is used to allow the initial randomly located pigs to group into a population that is distributed according to the rules of behavior captured in the model. Once the population stabilizes, control methods are applied.

8.3.4   Design Concepts

8.3.4.1   Recruitment

One of the challenges of controlling feral hogs is their rapid rate of reproduction. In this model, pigs are recruited into the population through birth. Because of the porous nature of the boundary of the area, the model uses a torus-shaped world that moves pigs exiting one edge to simultaneously enter into the study area at the opposite edge.

8.3.4.2   Weaning

Piglets remain attached to their mothers until they are weaned. At this time, females can become part of a sounder and males are forced to begin a more solitary life.

8.3.4.3   Attrition

Pigs die naturally at rates set within the model.

8.3.4.4   Social Grouping and Dynamics

After weaning, males begin a solitary life in which they hunt, establish territories, and attempt to mate with females in estrous. Females, on the other hand, form sounders and remain gregarious throughout their lives.

8.3.4.5 Diet

The location of swine on the landscape is based largely on the availability of food and water. While a pig's dietary requirements are extremely flexible, they are particularly attracted to certain foodstuffs, which change in abundance over the annual cycle.

8.3.4.6 Habitat, Range, and Travel

Sounders and solitary males maintain and defend home ranges that change with food and water availability and with the presence of competitors.

8.3.4.7 Control Methods

Two population control methods are provided in the model: hunting and the distribution of contraceptives in food baits.

8.3.5 Initialization

The model is initialized by (1) establishing the landscape characteristics based on GIS maps of the area, (2) randomly distributing a population of swine across parts of the landscape accessible to swine, and (3) ramping up the simulation by allowing the randomly located individuals to interact according to the behavior rules over the course of two simulation years.

8.3.6 Input

The model is built on a landscape consisting of a 206×213 grid of cells, with each cell representing a patch measuring 200×200 m. All input maps, including hog access, human access, mast (acorns and other tree nuts) production, and access to water, were derived from 2001 National Land Cover Data maps for the area (http://www.epa.gov/mrlc/nlcd-2001.html), and then resampled to this resolution. Hogs were allowed access to all areas except for dense urban areas and water bodies. Pigs were not killed by hunters in areas that are off-limits to humans.

Mast production is generally considered heavier in uplands, and lighter in bottomland woods (Stephen S. Ditchkoff, Auburn University, personal communication 2008). Mast abundance was predicted on the basis of relative local elevation and vegetative cover type. A patch was considered upland if it was in the top half of the elevation range within a 225-patch area. Patches were considered capable of mast production if classified as deciduous forest, mixed forest, woody wetlands, hardwood forest, xeric hardwood, xeric mixed pine-oak, mixed pine and hardwood, or bottomland hardwood. Each upland patch that was capable of mast production was assigned a mast potential value of 100, while those that were not were assigned a mast potential value of 0. The mast value applied to each 200×200-m patch in the model was the average of mast potential values within a 30-m, seven-patch diameter. At Fort Benning, the mast crop begins to fall in September; mast is present in our model from week 39 to week 13 and is absent at all other times.

For each patch, the distance to water was the minimum number of other patches a pig would have to enter to reach surface water. A patch was considered to contain surface water all year if classified as open water, cypress-gum swamp, or freshwater marsh, or if more than 500 other patches drained to it. September through February is Fort Benning's wet season, whereas March through August is considerably drier. For the dry season—model weeks 8–33—higher distance-to-water values were applied to patches.

8.3.7   Submodels

8.3.7.1   Recruitment

Feral swine can live at least 13 years, and no reproductive senescence has been noted. Female reproductive maturity can occur as early as 21 weeks (Hanson 2006), and may vary depending on nutrition and exposure to mature boars (males). While boars may be physiologically capable of breeding around 21 weeks of age, their fertility continues to increase over a period of several months (Flowers 2001). Competition is observed among boars attempting to breed (Pedersen 2007). In this model, gilts (young females) may farrow (give birth) at 43 weeks of age, which correlates to a minimum breeding age of about 26 weeks. Due to the inferior physiological fertility and competitive disadvantage of younger males, boars in this model begin breeding at 32 weeks.

While inadequate nutrition (Matschke 1964) and heat stress (Jolley 2007) have been shown to negatively affect male and female fertility, density-dependent decreases in fertility are not observed as long as the population is in good nutritional condition (Jolley 2007). Through necropsy of pregnant sows, feral pigs at Fort Benning have been found to carry as few as 3 to as many as 12 fetuses (Jolley 2007). The average litter for young females is significantly smaller than for older ones: at Fort Benning, females under 1 year old produce an average of 5 piglets, while older sows average 6.87 (Hanson 2006). In our simulation, females under 1 year old delivered 5 piglets. Sows 1 year or older had a 13% likelihood of delivering 6 piglets, and an 87% likelihood of delivering 7. Hanson (2006) observed a 1:1 sex ratio in the adult population; in our model, sex was randomly applied to each piglet as it was "born."

8.3.7.2 Farrowing

Feral piglets are weaned at 2–3 months of age (Choquenot et al. 1996), and sows may breed as soon as 1 week later. With a gestation of 114 days, the minimum farrowing interval for sows whose piglets survive to weaning is 26 weeks. Socialization with cycling sows and with boars induces estrus, so reproduction tends to be synchronized within sow groups (Pedersen 2007).

While domestic sows may farrow as many as three times in a 14-month period, such timetables are associated with managed weaning within days of birth. Feral piglets are unlikely to survive without maternal care at such an early stage. Under favorable conditions, feral sows are observed to produce approximately two litters annually. Breeding is considered nonseasonal (Jolley 2007), but timing is heavily dependent on food availability, especially during mast crop season (Graves 1984; Jolley 2007; Matschke 1964; Mauget et al. 1991). At Fort Benning, a bimodal breeding distribution is observed, with the majority of births occurring in March, and from July through November.

Because estrus tends to be synchronized within sow groups, semiannual farrowing schedules were assigned to sounders within the model. Farrowing dates were randomly determined based on a probability distribution that matched observed temporal birthing patterns.

8.3.7.3 Attrition

Feral pig mortality is high during the first 3 months of life (Hanson 2006). Depending on region, coyotes, bobcats, mountain lions, large raptors, and feral dogs are known to prey on feral swine, particularly young pigs. Little data are available on feral pig mortality prior to 1 month of age. Hanson (2006) provided survivorship data for Fort Benning pigs aged 1 month and older, but pigs less than 1 month old were too small to be trapped in that study. Pasture-raised domestic pigs may experience very high mortality during their first weeks of life; newborns can be crushed or suffocated by the sow, or succumb to hypothermia, starvation, disease, or predation (Honeyman and Roush 2002). For pigs 1 month old or less, we assigned survival rates of 42% based on preweaning mortality figures for pasture-raised domestic pigs.

In studies conducted at Fort Benning, very few adult feral pig deaths were reported due to predators, disease, starvation, or other nonanthropogenic causes. Humans are considered the primary predator for the species. When Hanson (2006) measured apparent survival rates on the installation, hunting was believed to cause approximately 90% of adult swine mortality. We derived annual mortality rates for pigs over 1 month of age directly from Hanson's apparent survival figures:

- Gilts 1–8 months old—0.311
- Gilts/sows over 8 months—0.319
- Boars 1–8 months old—0.200
- Boars over 8 months—0.207

To correspond with temporal and demographic partitions in the model, these rates were converted to weekly mortality for each age/sex class using the equation

$$DW = (1 - DA)^{1/52} - 1$$

where DW is the derived weekly morality and DA is the derived annual mortality rates for pigs over 1 month old. The resulting figures represent total attrition: emigration, natural mortality, and death caused by recreational hunting prior to the implementation of the bounty program.

8.3.7.4   Social Grouping and Dynamics

Sounders typically consist of 1–3 sows and their piglets, with an average size of 2.59 sows (Hanson 2006). When food is scarce, a sounder may become much larger. Male piglets typically remain with the sounder until they reach sexual maturity, at which time they may disperse widely (Hirotani and Nakatani 1987). Mature boars generally live and travel alone, but they congregate to follow sounders during estrus (Pedersen 2007). Mature females tend to remain much closer to their original home ranges (Hirotani and Nakatani 1987). Sounders in our simulation contained 2 or 3 mature sows and all of their immature offspring. Upon reaching sexual maturity, male offspring left the sounder. When a female reached sexual maturity, she was forced to leave her sounder if it already contained 3 older females. The expelled females then formed new sounders, with pigs in closer proximity more likely to end up in the same group. Each new sounder convened at the cluster of at least nine unoccupied patches nearest to a randomly selected member.

8.3.7.5   Diet

Movement and location of feral hogs appears to be driven largely by dietary requirements and food sources. Human land use and climate are also important factors in hog movement (Hanson and Karstad 1959). The diets of males and females do not significantly differ (Adkins and Harveson 2006). Roots, tubers, and herbaceous material comprise up to two-thirds of the summer diet, while the remainder includes mast, vertebrate and invertebrate fauna, bait, or other materials (Adkins and Harveson 2006; Taylor 1991). As food availability varies, swine may feed heavily on items such as earthworms, carrion, frogs, leeches, insects, eggs, small mammals, agricultural crops, and garbage (Graves 1984; Hanson and Karstad 1959; Herrero et al. 2006). Mast, consisting of hard-shelled seeds such as acorns and hickory, provides a source of high-quality nutrition because of its high fat and carbohydrate content (Graves 1984). During the winter when mast is available, it is the preferred food of feral swine (Adkins and Harveson 2006), comprising a majority of the diet (Graves 1984). Grass and forb intake increases in the summertime (Adkins and Harveson 2006). It is believed that grasses alone are insufficient to fulfill swine nutritional needs (Graves 1984), and an increase in destructive rooting behavior is observed as the availability of high quality above-ground food sources such as mast decreases (Graves 1984). Possibly due to their highly adaptive feeding behaviors, feral swine at Fort Benning do not appear to be approaching carrying capacity (Jolley 2007).

8.3.7.6 Habitat, Range, and Travel

Although cover can be essential for habitat selection depending on hunting pressure (Graves 1984), feral swine are observed in a broad range of habitats. At Fort Benning they are believed to occupy most habitat types, with the exception of urbanized areas. Our model excluded pigs from areas classified as urban or open water by National Land Cover Data.

Range data vary widely, with estimated seasonal ranges as small as 304 acres or as large as 6,175 acres (Graves 1984; Hanson and Karstad 1959). From approximately 1 month prior to birthing until piglets are about 4 weeks old, the sow moves away from the sounder (Hanson and Karstad 1959) and inhabits a much smaller area (Kurz and Marchint 1972). Boars tend to utilize larger home ranges than sounders (Graves 1984; Caley 1997), and may occupy up to twice as much terrain (Saunders and McLeod 1999). For both sexes, however, range size appears to depend on food and water availability (Hanson and Karstad 1959; Graves 1984). Feral swine exhibit strong fidelity to their home ranges (Graves 1984), generally leaving only when food (Kurz and Marchint 1972; Graves 1984) and/or water (Graves 1984; Taylor 1991) are inadequate, or to breed (Graves 1984), and returning when conditions are again suitable.

Sounder home ranges at Fort Benning are about 500–940 acres (Hanson 2006; Sparklin, unpublished data), and do not overlap (Sparklin, unpublished data). Swine are observed to move to wetland or bottomland areas within their ranges during the warmer, drier season, and into woodlands in the winter when the mast crop becomes available and water is more abundant. Simulated pig movement reflected these observations.

No published data were found for daily total distance walked or for net daily travel. One study found that the maximum distance between points visited on a given day was, on average, 0.4 miles for sounders and 0.7 miles for boars. These extremes in daily position depended on the availability of food and water (Kurz and Marchint 1972). Despite ample data regarding habitat preferences and range, the specific daily movements of feral pigs are considered difficult to predict (Stephen S. Ditchkoff, Auburn University, personal communication 2008).

The daily net distance traveled by modeled pigs was generated at random, and varied from zero to the average distances between extreme positions noted above. This randomly generated distance was applied to each boar and sounder at each time step, except when the constraints described below prevented a pig or group from moving to any adjacent patch. Ranges were not directly limited in size, shape, or environmental characteristics (except as stated above), but emerged as a result of the rules governing pig movement. The separation of sows from the sounder at farrowing was not modeled.

Lack of range overlap was achieved via an avoidance function. The model tracked which boar and which sounder had last visited each patch, and how long ago. This simulated "pig sign" allowed pigs to determine whether a patch was part of another pig's range, and it faded linearly over 52 weeks. The avoidance function prevented boars or sounders from entering patches visited within the last year by a pig of the same sex, unless they were otherwise unable to leave their patch.

Direction of travel was determined at each time step via a comparison of patches adjacent to a boar's or sounder's position. Comparisons were based on weighted values for distance to surface water, mast availability, and a random "attractiveness" factor. Distance to surface water was calculated as the distance to the nearest water. Distance-to-water and mast values each comprised 30% of direction-of-travel decisions. The remaining 40% was provided by a random number generated to account for other factors contributing to pig movement. Mirroring the behavior of feral swine at Fort Benning (Bill Sparklin, Montana State University, personal communication), simulated pigs moved seasonally among neighboring habitat types, establishing home ranges that provided surface water during the dry season as well as high-quality wintertime mast. At observed population sizes, boar ranges covered the majority of available land. Studies have found boar home ranges to be about twice the size of sounder ranges, but estimated Fort Benning population density and social groupings preclude this ratio. Because the number of boars is between two and three times the number of sounders, the study area is too small to allow for boar ranges larger than the typical area observed to be occupied by sounders at current densities.

8.3.7.7 Control Methods

All pigs occupying hunter-accessible areas were, while the hunt function was active, eligible to be hunted. In each step the model applies hunt-related mortality to randomly selected eligible pigs until either the specified number of hunt-related kills has been achieved or all eligible pigs have been killed.

Oral baits deliver contraceptives that indefinitely sterilize adult females. Techniques used in swine population control field trials could include ground placement by hand or aerial drop from a small aircraft (Fleming et al. 2000; Mitchell 1998), and bait feeding stations (Kavanaugh and Linhart 2000; Twigg et al. 2007). Nontarget species may remove a significant portion of the bait within the first few days following placement (Campbell et al. 2006; Fleming et al. 2000).

In our model, bait density was manipulated to emulate ground or aerial distribution techniques. We tested the distribution of different numbers of baits, both in single spatial blocks and by random dissemination between specified numbers of single patches. When the placement function was active, baits were distributed at user-specified intervals, with a default interval of 4 weeks. Baits remained in place for one time step, and only pigs in baited patches were eligible to ingest it. For every patch, the odds that each female pig would become sterile were ($b/2p$), where b is the number of baits in patch and p is the number of pigs in patch.

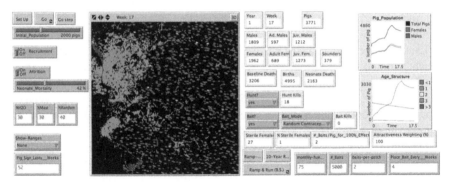

Fig. 8.2 Model interface

8.3.8 Interface

The NetLogo interface for this model is shown in Fig. 8.2. The darker objects (greenish when viewed in NetLogo) are user input controls. Adjustments to user-specified variables take effect immediately; the number of pigs killed by hunting and the quantity, frequency, and distribution of contraceptive bait can be modified while an experiment is running. The lighter objects (light brown when viewed in NetLogo) show system state information. In this model, the pig population counter and the graphs showing age and sex demographics are updated at the end of each time step. The three buttons at the top left initialize and run the model, and the large square in the center shows the mapped state of the system, including the actual locations of pigs on the installation. Because groups of up to ten sows and their piglets may move together, the total pig population may be significantly greater than the number of pig icons visible.

8.4 Simulation Experiments

The model was used to test all 16 combinations of 4 hunt and 4 contraceptive scenarios. The hunt levels were 0, 25, 50, and 75 kills per month. The bait levels were 0, 2,500, 5,000, and 7,500 baits placed per month. Each scenario was simulated ten times, and ended when either 12 model years had passed or the simulated population exceeded 12,000 pigs. In each simulation our modeled population quickly achieved an age distribution similar to that seen in the literature, with only a small percentage of pigs surviving past 2 years of age. Each scenario began with 1,100 swine generating a population after a 104-week "ramp-up" during which there was no hunting or baiting. Population growth rates generally remained within the measured range, consistently approaching the 142% mean observed at Fort Benning (Hanson 2006).

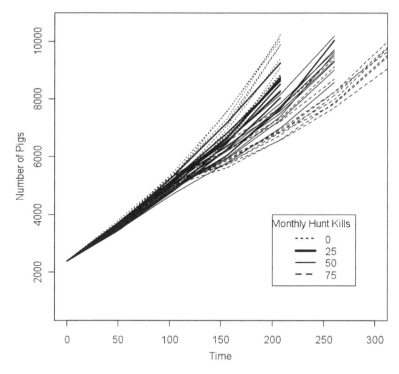

Fig. 8.3 Various hunting intensity, no contraceptive bait

Without the use of contraceptive baits, even high-intensity hunting ultimately had little impact on the size of the feral pig populations although it did slightly reduce the rate of population growth (Fig. 8.3). In these figures, each trace represents 1 of the 10 trials for that scenario. When used in the absence of hunting, contraceptive baits were similarly ineffective at low intensity (Fig. 8.4). However, when a larger number of baits were placed each month, contraception showed potential for limiting the size of the feral pig population.

With low-intensity contraceptive use, population growth slowed but did not stop as the number of pigs killed by hunting was increased (Fig. 8.5). In the inverse scenario, with hunting kept at low intensity, growth rates were similarly responsive to increases in contraceptive baiting (Fig. 8.6). Population control was consistently achieved at higher hunting intensities in these lower-intensity contraception scenarios.

For any time step in a simulation, variation in the pig population across treatments was inversely related to treatment intensity. For simulations in which treatment intensity was adequate to prevent the pig population from reaching 12,000 individuals, greater differences between treatments were seen at the 8- and 12-year time points. Figure 8.7 shows how the population at the 8-year mark (416 weeks) varied with kill rate, represented by individual lines showing the average population across replicates, and contraceptive bait intensities.

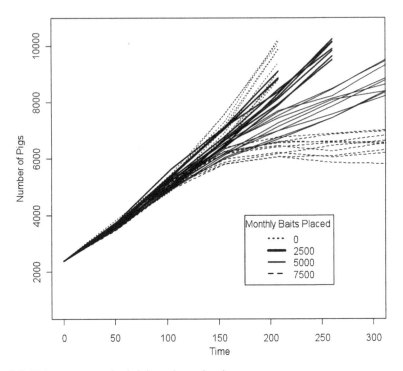

Fig. 8.4 Various contraceptive bait intensity, no hunting

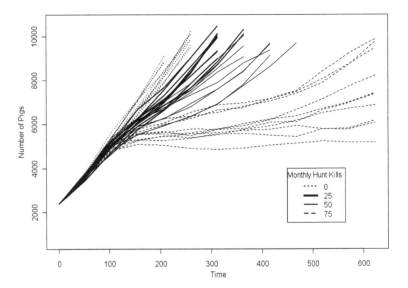

Fig. 8.5 Low contraceptive bait intensity, various hunt intensities

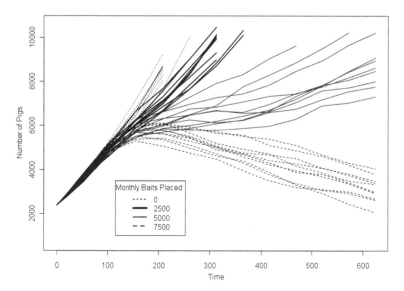

Fig. 8.6 Low hunt intensity, various contraceptive bait intensities

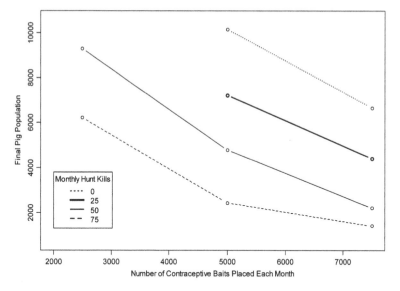

Fig. 8.7 Average population at week 416 for different hunt rates across various bait intensities

When moderate to high-intensity hunting was combined with contraceptive baiting, population reduction and control were consistently achieved. Population control was feasible with low, moderate, or high-intensity placement of contraceptive baits in combination with moderate or high-intensity hunting. Figure 8.8 shows population changes using monthly totals of 25 kills with 7,500 baits, 50 kills with 5,000 baits, and 75 kills with 2,500 baits.

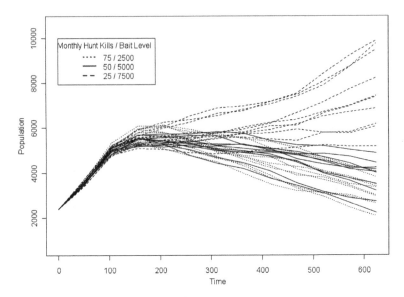

Fig. 8.8 Combined moderate hunting and contraceptive baiting intensities

8.5   Discussion

Age distribution, population growth rates, and seasonal movement produced by our model were consistent with general data available for feral swine and with observations specific to the Fort Benning population. Our results support the hypothesis that the combination of lethal control and oral contraceptive delivery will provide better control of the Fort Benning feral swine population than will either technique alone. We did not evaluate the impacts of trapping, bait stations, bait-and-euthanize schemes, or the sterilization of males. Future modeling studies would be useful to determine whether and how these techniques might be included in an integrated management program.

Behavioral adaptations, such as aversion to hunting and trapping sites or increased frequency of visits to baited areas, have been reported, and could warrant specific attention in the development of long-term control strategies. Variation in surface water availability due to annual climate variation or climate change could also affect feral pig ranges. Future modeling efforts should investigate the potential impact of these factors on feral pig movement and management.

8.6   Conclusions

Our study suggests that a combination of lethal and contraceptive techniques is more likely than either method alone to achieve management goals for the Fort Benning feral swine population. Improved control of this population by combined methods is likely to reduce habitat destruction and disease risks to threatened and endangered species, and improve overall achievement of conservation management objectives. In addition, our model provides a framework for understanding how feral swine interact with the landscape and helps management decision-makers predict the effects of proposed control techniques on swine populations and their location. The ability to test such controls could improve cost- and labor-efficiency of invasive species management, particularly as control options are reevaluated in the context of new information, alternative management scenarios, or changing conditions.

References

Adkins RN, Harveson LA (2006) Summer diets of feral hogs in the Davis Mountains, Texas. Southwest Nat 51(4):578–580

Caley P (1997) Movements, activity patterns and habitat use of feral pigs (*Sus scrofa*) in a tropical habitat. Wildl Res 24:77–87

Campbell TA, Lapidge SJ, Long DB (2006) Using baits to deliver pharmaceuticals to feral swine in southern Texas. Wildl Soc Bull 34(4):1184–1189

Choquenot D, McIlroy J, Korn T (1996) Managing vertebrate pests: feral pigs. Bureau of resource sciences. Australian Government Publishing Service, Canberra

Dzieciolowski RM, Clarke CMH, Frampton M (1992) Reproductive characteristics of wild pigs in New Zealand. Acta Theriologica 37:259–270

Fleming PJS, Choquenot D, Mason RJ (2000) Aerial baiting of feral pigs (*Sus scrofa*) for the control of exotic disease in the semi-arid rangelands of New South Wales. Wildl Res 27(5):531–537

Flowers WL (2001) Effect of age at which semen collection regimens are initiated on production of spermatozoa in boars. North Carolina State University, College of Agriculture and Life Sciences, Department of Animal Science Annual Swine Report 2001

Graves HB (1984) Behavior and ecology of wild and feral swine (*Sus scrofa*). J Anim Sci 58(2):482–492

Hanson LB (2006) Demography of feral pig populations at Fort Benning, Georgia. Auburn University, Auburn

Hanson RP, Karstad L (1959) Feral swine in the southeastern United States. J Wildl Manage 23(1):64–74

Herrero J, Garcia-Serrano A, Couto S, Ortuno VM, Garcia-Gonzalez R (2006) Diet of wild boar *Sus scrofa* L. and crop damage in an intensive agroecosystem. Eur J Wildl Res 52(4):245–250

Hirotani A, Nakatani J (1987) Grouping-patterns and inter-group relationships of Japanese wild boars (*Sus scrofa* leucomystax) in the Rokko mountain area. Ecol Res 2:77–84

Honeyman MS, Roush WB (2002) The effects of outdoor farrowing hut type on prewean piglet mortality in Iowa. Am J Altern Agric 17(2):92–95

Jolley DB (2007) Reproduction and herpetofauna depredation of feral pigs at Fort Benning, Georgia. Auburn University, Auburn

Kavanaugh DM, Linhart SB (2000) A modified bait for oral delivery of biological agents to raccoons and feral swine. J Wildl Dis 36(1):86–91

Kurz JC, Marchint RI (1972) Radiotelemetry studies of feral hogs in South Carolina. J Wildl Manage 36(4):1240–1248

Matschke GH (1964) The influence of oak mast on European wild hog reproduction. In: Proceedings Annual Conference Southeast Association of Game and Fish Commission, vol 18. pp 35–39

Mauget R (1991) Reproductive biology of the wild Suidae. In: Barrett RH, Spitz F (eds) Biology of Suidae. IRGM, Toulouse, pp 49–64

Mitchell J (1998) The effectiveness of aerial baiting for control of feral pigs (*Sus scrofa*) in north Queensland. Wildl Res 25(3):297–303

Pedersen LJ (2007) Sexual behaviour in female pigs. Horm Behav 52(1):64–69

Saunders G, McLeod S (1999) Predicting home range size from the body mass or population densities of feral pigs, *Sus scrofa* (Artiodactyla: Suidae). Aust J Ecol 24:538–543

Taylor RB (1991) The feral hog in Texas. In: Federal Aid Report Series Number 28. Texas Parks and Wildlife Department, Austin

Twigg LE, Lowe T, Martin G (2007) Bait consumption by, and 1080-based control of, feral pigs in the Mediterranean climatic region of south-western Australia. Wildl Res 34(2):125–139

Wilensky U (1999) NetLogo. Computer software. Center for Connected Learning and Computer-Based Modeling, Northwestern University, Evanston. http://ccl.northwestern.edu/netlogo/

Chapter 9
Spatially Explicit Modeling of Productivity in Pool 5 of the Mississippi River

Katherine R. Amato, Benjamin Martin, Aloah Pope, Charles Theiling, Kevin Landwehr, Jon Petersen, Brian Ickes, Jeffrey Houser, Yao Yin, Bruce Hannon, and Richard Sparks

9.1 Background

Understanding what fuels large river ecosystems is important because rivers produce, utilize, store, and transport energy in the form of organic carbon, thereby playing an important role in the global carbon cycle (Shih et al. 2010). Virtually all large rivers in the world have been developed for water supply, navigation, and flood control, but at a cost to valuable natural goods (e.g., fish) and services (e.g., sustainment of biodiversity) that depend on natural sources of energy (Vorosmarty et al. 2010). However, it is difficult to determine what fuels large river ecosystems because of the spatial and temporal complexity of floodplain-river ecosystems. The flow of organic carbon in river ecosystems is controlled not only by biotic and abiotic factors similar

K.R. Amato (✉)
Program in Ecology, Evolution, and Conservation Biology, University of Illinois,
286 Morrill Hall, 505 South Goodwin Ave, Urbana, IL 61801, USA
e-mail: amato1@illinois.edu

B. Martin • A. Pope
Department of Natural Resources and Environmental Sciences, University of Illinois,
W-503 Turner Hall, 1102 South Goodwin Ave, Urbana, IL 61801, USA

C. Theiling • K. Landwehr • J. Petersen
U.S. Army Corps of Engineers, 1500 Rock Island Dr, Rock Island, IL 61201, USA

B. Ickes • J. Houser • Y. Yin
Upper Midwest Environmental Sciences Center, U.S. Geological Survey,
2630 Fanta Reed Road, La Crosse, WI 54603, USA

B. Hannon
Department of Geography, University of Illinois, 220 Davenport Hall,
607 S Mathews, M/C 150, Urbana, IL 61801, USA

R. Sparks
Illinois Natural History Survey, University of Illinois,
P.O. Box 176, Elsah, IL 62028, USA

J.D. Westervelt and G.L. Cohen (eds.), *Ecologist-Developed Spatially Explicit Dynamic Landscape Models*, Modeling Dynamic Systems,
DOI 10.1007/978-1-4614-1257-1_9, © Springer Science+Business Media, LLC 2012

to those observed in terrestrial ecosystems (e.g., "who eats whom" and ambient temperature, respectively) but also by hydraulic factors that move carbon within the river system (Doi 2009; Power 2006).

It is easy to understand that the ultimate source of energy is the carbon fixed by green plants through photosynthesis, but these primary producers (called *autotrophs*, literally meaning "self-feeders") live in complex distribution patterns along tributary streams, on vast floodplains, and in the river itself in the form of microscopic algae and rooted or floating *macrophytes*. The consumers of carbon (i.e., *heterotrophs*) include mobile fauna such as fish and zooplankton, and relatively immobile animals such as mollusks, worms, and aquatic insects that attach to or burrow into the bed of the river. Additionally, there are the decomposers such as bacteria and fungi, which utilize the waste products or dead bodies of both the producers and consumers.

In rivers, as in most other ecosystems, carbon does not move in a simple, linear food chain from plants to herbivores to smaller and larger predators. Instead, consumers such as zooplankton, fish, and mollusks may use several sources of organic carbon, including algae, particulate organic carbon (POC), and other smaller consumers. The microbial layer of bacteria, fungi, and algae that coats the river bed and solid objects in the river (e.g., mollusk shells, sunken logs, and built structures) can take up dissolved organic carbon (DOC). The resulting feeding linkages look less like a food chain and more like a *food web*.

The complexity of the food web is compounded by the fact that the distribution of producers and consumers varies both spatially and temporally. Annual cycles in day length, temperature, water depth, and water flow typically distinguish a warm season of reproduction and growth from a cold season of relative inactivity. Because the dynamics of riverine food webs are so complex, it is not surprising that there is considerable debate about what controls productivity in large rivers (Dettmers et al. 2001; Hoeinghaus et al. 2007; Power 2006; Power and Dietrich 2002) and that there are alternative riverine productivity theories.

Three current theories that attempt to explain patterns of productivity in large rivers are (1) the river continuum concept, or RCC; (2) the riverine productivity model, or RPM; and (3) the flood-pulse concept, or FPC (Dettmers et al. 2001). The RCC states that leaves from terrestrial plants that fall into headwater streams are the main source of carbon, which is assimilated and transferred downstream with water flow (Vannote et al. 1980). The RPM states that plants growing in the river, including microscopic algae and rooted macrophytes, generate most of the carbon consumed by crustaceans, mollusks, aquatic worms, aquatic insects, and fish (Thorp and Delong 1994). The FPC asserts that consumers in the river depend mostly on carbon produced in the adjacent floodplains by aquatic or flood-tolerant plants. Fish and other mobile consumers can access the resources of the floodplains during seasonal floods, and carbon produced on the floodplains can also be carried into the river when the floods recede (Junk et al. 1989). The FPC regards the seasonal flood pulse as the major driving force responsible for the existence, interactions, and productivity of the major biota in large floodplain-river ecosystems.

Field data have provided some support for each of these three theories across different river systems and within river systems across seasons (Gutreuter et al. 1999;

Hoeinghaus et al. 2007; Kohler 1995; Oliver and Merrick 2006; Schramm and Eggleton 2006; Sheldon and Thoms 2006). However, no single theory applies uniformly across time or space, and in many cases it remains unclear whether carbon produced in the river (*autochthonous* carbon) or outside of the river (*allochthonous* carbon) is the most important driver of large river productivity. One possibility is that all three sources of productivity are important, but in different places at different times of year (e.g., carbon from upstream and from floodplains adjacent to the river may be important during seasonal floods, but not during the low flow season).

The question of what fuels river ecosystems is not merely of theoretical interest because the answer could determine priorities for managing and restoring rivers. If fish and other organisms valued by humans depend on upstream sources of carbon, then an efficient restoration strategy would be to work from the headwaters down. However, if riverine consumers obtain their carbon mostly from the river itself, then from an energy standpoint it matters little whether the floodplains are periodically connected to the river, although there may be other reasons for maintaining or restoring river-floodplain connectivity, such as providing flood storage (Akanabi et al. 1999).

Built structures such as levees, locks, and dams may alter natural patterns of productivity by altering water depth, flood dynamics, and sediment accumulation (Tyser et al. 2001). On the Upper Mississippi River, there are 29 navigation dams and locks in place from Minneapolis, Minnesota, to St. Louis, Missouri. Downstream of St. Louis, the river is free-flowing to the Gulf of Mexico. Most of the floodplain in Wisconsin and Minnesota is unleveed and connected to the river during seasonal floods, while approximately half of the floodplain in Iowa, Illinois, and Missouri is leveed, mostly for agriculture purposes (Delaney and Craig 1997). Despite the presence of dams and levees, the Upper Mississippi River retains a complex lateral and upstream-downstream mosaic of landforms and ecological communities, and exhibits processes characteristic of large floodplain-river ecosystems (Sparks et al. 1990). Typically, each reach of the river that is defined by navigation dams, termed a *navigation pool* by the US Army Corps of Engineers, consists of a downstream portion that is impounded during the seasonal low flow to maintain water depths for navigation (Fig. 9.1). During moderate seasonal floods, the dam gates are raised from the bottom of the river and out of the water. During major floods, the dams (typically low earthen weirs that connect the gates to the shore) are overtopped. Even during low flows, the influence of the dams extends only approximately half the distance upstream to the next dam. Upstream of the halfway point, the river retains the meandering main channel, numerous side channels, islands, backwaters, and floodplains that characterized the original natural environment (Fig. 9.1). The navigation dams have, in effect, created multiple replicate systems that are well suited for hypothesis testing. The subject of this study was Pool 5, for which a hydraulic model and biological data were available.

Spatially explicit modeling of large river ecosystems offers a study approach that enables researchers to use both short-term and long-term data to develop restoration and management plans. Field studies of food webs are expensive. The results are usually valid only for a few years, and at best they provide snapshots of river dynamics that may not represent either prevailing trends across decades or long-term responses to significant environmental disturbances. By contrast, long-term monitoring programs

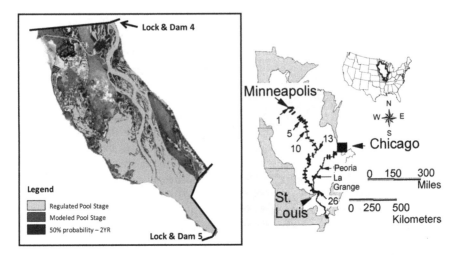

Fig. 9.1 Map of Pool 5 on the Upper Mississippi River. The darkest colors represent sections of the river that have a 50% probability of being submerged during any given year. Two years are predicted to pass before these sections will be submerged

may detect trends in populations and productivity over time, but they fail to identify the underlying causes of the trends. Models can be used to determine whether the spatial and temporal patterns of primary production and consumption identified in short-term studies play a role in long-term population trends detected by monitoring, and they also can efficiently identify gaps in current databases to guide future data collection. Although modeling is an excellent tool for planning and directing long-term productivity studies, the creation of a riverine model can be time-consuming and expensive. Furthermore, while numerous studies have used modeling to examine river ecosystem dynamics in terms of productivity (Best and Boyd 2008; Best et al. 2001; Garbey et al. 2006; Herb and Stefan 2003; Park et al. 2003), many of these models are limited in their ability to incorporate the physical complexities of river ecosystems. This chapter describes a basic river productivity simulation model that builds off of previously constructed models and also incorporates ecological and hydraulic processes, thereby improving overall model realism while reducing the time and cost requirements of making those improvements.

9.2 Objective

The objective of this project was to integrate river hydrology into an ecological model of river productivity. By uniting ecological and hydraulic processes within the same model we seek to better represent the complexity of the Mississippi River carbon cycle and to pinpoint key sources of productivity within it. If our model of the food web within the Mississippi River can accurately predict productivity in the river, it will support the RPM (in-river production of organic carbon is sufficient to fuel the river ecosystem); if this is not the case, then organic carbon from upstream sources, the floodplain, or both may also play important roles and should be incorporated into future models.

A secondary objective of this work was to design the model so it can easily be applied to a variety of river ecosystems simply by changing hydrology inputs such as maps of depth, velocity, and flow direction. By meeting the secondary objective, this model may provide useful guidance for management planning by addressing variations in hydrology from locks and dams as well as natural processes such as sedimentation.

9.3   Model Description

9.3.1   Purpose

This model was developed using NetLogo 4.0.4 (Wilensky 1999).[1] The purpose of the model is to combine ecological and hydrological processes to simulate carbon flow through Pool 5 of the Upper Mississippi River, a 24 km section extending from Navigation Dam No. 4 at Alma, Wisconsin, downstream to Navigation Dam No. 5 near Whitman, Minnesota (Fig. 9.1). A basic carbon cycle was modeled using simple food web interactions. Water depth, current velocity, and current direction were derived for Pool 5, assuming a moderate discharge of 88,000 cfs, using the Adaptive Hydraulics 2D hydraulic simulation model (USACE 2008) to describe the spatial distribution of dissolved carbon.

9.3.2   State Variables and Scales

Water depth, velocity, and flow direction in Pool 5, as reported by the US Army Corps of Engineers, were plotted on digital maps using a color gradient. These maps were divided into patches of 30×30 m (Fig. 9.2), and the carbon cycle was modeled within each patch (as described in Sect. 9.3.6). This carbon cycle model, originally developed in 1983 for Pool 19 of the Upper Mississippi River (Sparks 1985), simulates the transfer of labile carbon among trophic levels within each patch. It includes 11 basic carbon states: DOC, POC, detritus, phytoplankton, macrophytes, periphyton, herbivores, decomposers in the water column, decomposers in the sediment, consumers in the water column, and consumers in the sediments. The abundance of each of these stocks was expressed in grams of carbon per patch.

---

[1] An operational copy of this model is available through http://extras.springer.com.

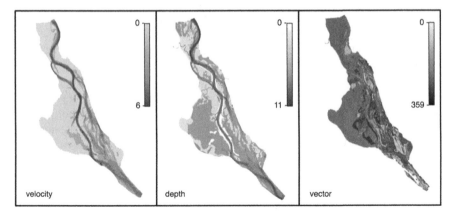

Fig. 9.2 Grayscale GIS maps of water velocity, depth, and flow direction in Pool 5. *Darker* colors represent increased velocity and depth, as well as changes in the direction of flow. Maps were created using data from a US Army Corps of Engineers ADH 2D hydraulic model. Units are m/s/30 m² (velocity), m/30 m² (depth), and degrees/30 m² (vector, from 0 to 360°)

While there is evidence that primary productivity in large river ecosystems is influenced by nitrogen, phosphorous, and other nutrients (Allan and Castillo 2007; Francoeur 2001; Houser and Richardson 2010), there is other evidence that the Upper Mississippi River is not nutrient-limited and that primary production, especially in the main channel, is often limited by factors such as light availability and temperature (Houser et al. 2010; Huff 1986; Owens and Crumpton 1995). For these reasons, and to maintain simplicity, the effects of nitrogen and phosphorous are not simulated in this model. Primary productivity is based on temperature, light, and water velocity. Transfer of carbon from one stock to another is based on biomass-dependent interactions between stock. The movement of labile carbon between patches is modeled on the basis of water velocity and flow direction (see Sect. 9.3.6.4). The time step for the model is set at 1 h for the purpose of corresponding with many of the measurements reported for river processes in the literature and with the hydrology data.

9.3.3   Process Overview and Scheduling

Within each patch, during each time step, carbon is first added to both autotrophic and heterotrophic carbon stocks as dictated by the formulas for primary production and consumption. Then the carbon that was consumed by the heterotrophs is removed from the appropriate carbon stocks indicated by the food web. Finally, each stock is assigned a new biomass value that takes growth and predation into account, and the hydrologic component moves carbon among patches. Interactions between stocks are described below.

9.3.4 Initialization

All carbon stocks transported in the current (phytoplankton, water column decomposers, DOC, POC) were arbitrarily set at 10 g of carbon per patch to start the model. These stocks also received influxes of 10 g of carbon in each of the patches bordering the upstream end of Pool 5 during each time step to simulate the influence of upstream carbon on the system. This amount was chosen based on organic suspended sediments data for Pool 4 from the Long-Term Resource Monitoring Program implemented by the US Geological Survey (http://www.umesc.usgs.gov/ltrmp.html). All other stocks were set at 1 g of carbon per cell to start the model and were not subject to influxes or outfluxes during the simulation.

9.3.5 Input

The model takes input from GIS maps of water depth, velocity, and flow direction in Pool 5 of the Mississippi River, which was created from a 2D hydraulic model developed by the US Army Corps of Engineers (USACE 2008). The hydraulic model simulated a discharge of 2,097 m^3s^{-1}. Although this discharge level is not common for periods longer than 1 month, it was chosen because the data from the hydraulic model were the most readily available source. Maps were divided into patches across a color gradient so that an average depth, velocity, and flow vector could be calculated for each patch.

9.3.6 Submodels

9.3.6.1 Autotrophic Stocks

The two autotrophs considered for this model are macrophytes and phytoplankton. Macrophytes were modeled only in patches where water velocity was less than 1 m/s because higher velocities result in complete scouring.

We modeled all autotrophic stocks using the general equation:

$$dX_j / dt = \mathrm{Prod}_j - \mathrm{Phys}_j - \mathrm{Pred}_j \qquad (9.1)$$

where the biomass of stock j equals the difference between primary production and physiological and predatory losses.

Primary production for macrophytes is described using the maximum gross photosynthesis equation (Chambers et al. 1991; Sand-Jensen et al. 2007):

$$\mathrm{Prod}_j = 0.08 \times X_j (l / (l+10))(K - X_j) / K \qquad (9.2)$$

where X_j is macrophyte biomass, K is the half-saturation constant, and l is the amount of light that reaches the macrophytes depending on the amount of surface light, depth, and total suspended solids:

$$l = (se^{((-d)b)})$$ (9.3)

where s is surface light, d is depth, and b is turbidity. Primary production for phytoplankton was modeled using a formula similar to that used for macrophytes (Huisman and Weissing 1994).

Physiological losses of an autotrophic stock were modeled as:

$$\text{Phys}_j = (\mu_j + \eta_j + \sigma_j)X_j$$ (9.4)

where μ_j is the specific physiological mortality rate, η_j is carbon lost to excretion, and σ_j is carbon lost to respiration (Garbey et al. 2006). Carbon lost via physiological mortality is transferred to the detritus, POC, and DOC stocks. Carbon lost via excretion enters the DOC stock. Carbon lost to respiration by a stock is removed from the system. In this version of the model, the phytoplankton stock—not the macrophyte stock—loses carbon to herbivores. These losses are described in the equations for heterotrophic stocks.

9.3.6.2 Heterotrophic Stocks

Heterotrophic stocks included aufwuchs (microbial component), herbivores, consumers, and decomposers. We modeled all of these stocks using the general equations developed by Wiegert (1975). Biomass of a stock at time $t+1$ is determined by the formula:

$$dX_j / dt = C_j(1 - \varepsilon_j) - \text{Phys}_j - \text{Pred}_j$$ (9.5)

where the biomass of stock j equals the difference between consumption corrected for egestion (ε_j) and physiological and predatory losses. Egested material enters the detrital food web as POC, much of which may be converted to detritus depending on the hydrological conditions of the cell. Consumption (C) by stock j is determined by the summation of consumption of each available prey type (m) of predator j, given by the formula:

$$C_j = \sum (\pi_{ij}\tau_j X_j f_{ij} f_{jj}) \quad \left(\sum \text{ is sum from } i = 1 \text{ to } m \right)$$ (9.6)

where π_{ij} is the preference of stock j for prey stock i, τ_j is the maximum rate of consumption by predator stock j, and X_j is the biomass of predator stock j. The functional response of predator stocks is determined by f_{ij} and f_{jj}, which represent prey and space limitation functions:

$$f_{ij} = y / (A_{ixy} - G_{ixy})$$ (9.7)

$$f_{jj} = 1 - ((x - A_{jx}) / (G_{jx} - A_{jx}))$$ (9.8)

Table 9.1 Stock-specific values for prey limitation of heterotrophs

Predator stock	Prey stock	π_{ij}	A_{ij}	G_{ij}
Herbivore	Phytoplankton	0.7	20	0.01
Herbivore	Periphyton	0.15	3	0.01
Herbivore	Water decomposers	0.15	3	0.01
Water decomposer	POC	0.5	30	0.05
Water decomposer	DOC	0.05	30	0.05
Sediment decomposer	Detritus	1	0.6	0.005
Periphyton	DOC	0.5	30	0.05
Periphyton	POC	0.5	30	0.05
Sediment consumer	Sediment decomposer	0.5	3	0.02
Sediment consumer	Detritus	0.1	3.5	0.02
Sediment consumer	Periphyton	0.4	2	0.02
Consumer	Herbivore	0.7	3.5	0.025
Consumer	Sediment consumer	0.3	4	0.04

where x is the present biomass of the heterotrophic stock in question and y is the present biomass of the prey stock being fed on by the heterotrophic stock. Because population growth rates can be limited by both prey availability and intraspecific interference at high population densities, both f_{ij} and f_{jj} produce values between 0 and 1. At high prey densities, predators will not be resource-limited and thus will feed at the maximum rate (τ_j). However, below a prey density threshold (A_{ij}), consumption will be reduced and will eventually fall to zero when prey densities are below a refuge level (G_{ij}). If a predator stock feeds only on one prey type, then π_{ij} is set to 1. However, when a predator stock feeds on multiple stocks, π_{ij} is a function of innate prey preference (P_{ij}) for specific stocks and the biomass of those stocks in the cell, given as:

$$\pi_{ij} = (P_{ij} f_{ij}) / \sum (P_{ij} f_{ij}) \quad \left(\sum \text{is sum from } i = 1 \text{ to } m\right) \qquad (9.9)$$

For simplicity we assume f_{ij} decreases linearly between A_{ij} and G_{ij}. Space limitation is determined similarly, except values of A_{jj} and G_{jj} relate to predator density.

Stock-specific values for A_{ij}, A_{jj}, G_{ij}, G_{jj}, and P_{ij} were adapted from Pace (1984) and are listed in Tables 9.1 and 9.2. However, these values are likely to be system-specific and thus should be calibrated with independent data from the system to be modeled. Calibration for this system will occur with further model validation and sensitivity analyses.

Physiological losses of a heterotrophic stock were modeled as above for autotrophic stocks:

$$\text{Phys}_j = (\mu_j + \eta_j + \sigma_j) X_j \qquad (9.10)$$

Carbon lost to predation is determined by the summation of losses to each predator stock (k).

$$\text{Pred}_j = \sum (\pi_{jk} \tau_k X_k f_{jk} f_{kk}) \quad \left(\sum \text{is sum from } k = 1 \text{ to } n\right) \qquad (9.11)$$

Table 9.2 Stock-specific values for space limitation of heterotrophs

Stock	T	P	H	μ	ε	A_{ij}	G_{ij}
Herbivore	0.04166667	0.003333	0.016666667	0.000416667	0.2	2.4	108
Water decomposer	0.26083333	0.025	0.007208333	0.002083333	0	1.2	20
Sediment decomposer	0.26083333	0.025	0.007208333	0.002083333	0	0.2	120
Periphyton	0.08333333	0.0025	0.004166667	0.000416667	0.4	1% of macro	20% of macro
Sediment consumer	0.02291667	0.003333	0.000416667	0.000416667	0.35	2	20
Consumer	0.005125	0.000521	8.33333E-05	8.33333E-05	0.2	0.65	6.5

Table 9.3 Percent of senesced, excreted, and egested carbon assigned to new carbon states

Carbon source	Detritus	DOC	POC
Senescence			
Macrophytes	100	0	0
Scoured macrophytes	90	1	9
Phytoplankton	90	1	9
Herbivore	30	0	70
Periphyton	70	0	30
Water column consumer	100	0	0
Sediment consumer	100	0	0
Water column decomposer	30	0	70
Sediment decomposer	100	0	0
Excretion	0	100	0
Macrophyte exudation	0	100 (4% of carbon from photosynthesis)	0
Egestion	100	0	0

9.3.6.3 Detritus, POC, and DOC

Both the detritus and POC carbon stocks acquire carbon from dead organic matter that is greater than 1 μm in size. However, POC refers to matter in the water column and detritus refers to matter in the sediments. For this model, senesced material was deemed detritus if particle size was greater than 10 μm (Wetzel 2001). The percent of senesced material transferred to the detritus stock varied according to its source (Table 9.3). All egested carbon contributes to detritus (Table 9.3). Carbon lost to POC from the detritus category is described as:

$$\text{Trans}_j = X_j(2.5\log((v/40.0)+0.0001)+0.5) \tag{9.12}$$

where v is equal to the water velocity, and higher velocities result in a greater proportion of organic matter in the water column. Similarly, the transfer of carbon from the POC stock to the detritus stock is described as the product of POC biomass and $(1-\text{Trans}_j)$. When velocity is equal to zero, 90% of the carbon in the POC stock settles out and becomes detritus.

All senesced carbon in the model that is not assigned to the detritus stock is assigned to the POC stock, with the exception of a small percentage that is assigned to the DOC stock (Table 9.3). Carbon is also transferred to POC from DOC at a rate of 1% due to flocculation.

The DOC carbon stock acquires carbon from dead or excreted objects that measure less than 1 μm in size (Wetzel 2001). In this model, in addition to some senesced carbon, all carbon excreted or released as exudates contributes to DOC, seen in Table 9.3 (Wetzel 1984).

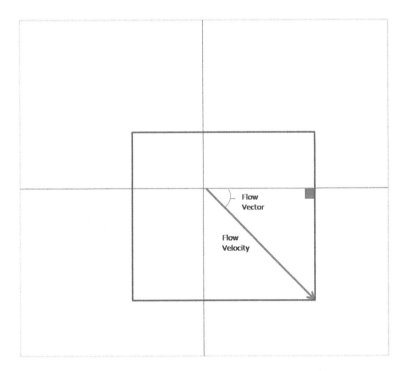

Fig. 9.3 Trigonometric design of the hydrological component of the model

9.3.6.4 Incorporation of Hydraulics

Mobile carbon from each cell was modeled to flow at the same velocity as water, and the amount of carbon transferred between cells was calculated using trigonometric functions based on the speed and direction of water flow in each cell (Fig. 9.3). Since most fish actively move to find food, consumers in the water column in this model were not affected by hydraulics.

9.4 Simulation Experiments

9.4.1 Model Run

The model was run for 1,368 time steps, which is equivalent to approximately 57 days of simulated time. The simulation took approximately 1 week to complete running on a desktop computer. The starting values for each stock and carbon influxes into the pool are listed in Tables 9.4 and 9.5. The amount of carbon (g) assigned to each carbon stock was recorded at each time step for the duration of the simulation,

Table 9.4 Model starting values for carbon stocks

Stock	Starting value (g C/cell)
Macrophytes	1.0
Phytoplankton	10.0
Herbivores	1.0
Water column consumers	0.1
Sediment consumers	1.0
Water column decomposers	10.0
Sediment decomposers	1.0
Detritus	1.0
POC	10.0
DOC	10.0

Table 9.5 Values for influxes at the top of the navigation pool (upstream end)

Stock	Influx value (g C/cell)
Phytoplankton	10.0
Water column decomposers	10.0
POC	10.0
DOC	10.0

Values for outfluxes at the bottom of the pool (downstream end) were equal to influx values

and changes in each carbon stock were plotted as a function of time. The final size of each carbon stock in each cell was also recorded. These data were translated into maps representing each stock.

9.4.2 Simulation Results

The amount of carbon in each stock remained relatively constant throughout the simulation or plateaued after an initial increase. The patches bordering the downstream end of the pool had measured outfluxes of 10 g of carbon during each time step, which was equal to the programmed influx into the patches bordering the upstream end. The greatest amount of carbon biomass accumulated in areas of moderate depth and velocity directly adjacent to the main channel (see Figs. 9.4–9.7). Detritus accumulated in most of the pool during the 57-day simulation, which is not surprising since the relatively low discharge promotes the settling of organic matter in low-current, off-channel areas (Fig. 9.6). The patterns for individual stocks are detailed below.

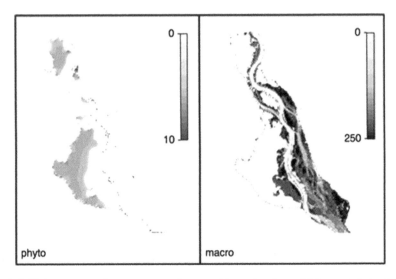

Fig. 9.4 NetLogo grayscale map output of autotrophic stocks after a 57-day model simulation. *Darker* colors represent higher carbon biomass. Units are grams of carbon/30 m²

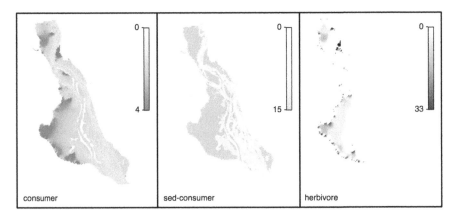

Fig. 9.5 NetLogo grayscale map output of heterotrophic consumer stocks after a 57-day model simulation. *Darker* colors represent higher carbon biomass. Units are grams of carbon/30 m²

9.4.2.1 Autotrophic Stocks

Primary producers were present throughout Pool 5 (Fig. 9.4). Therefore, although the amount of carbon available to consumers from producers varied depending on location in the pool, producers were available for consumers in all parts of the pool. Macrophytes and phytoplankton did not coexist in any portion of the pool. Phytoplankton biomass was greatest in several off-channel areas, including the

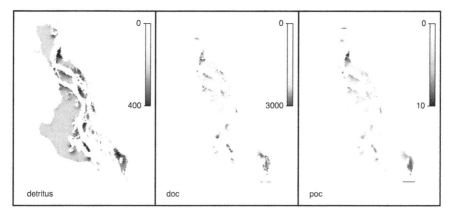

Fig. 9.6 NetLogo grayscale map output of detritus, DOC, and POC after a 57-day model simulation. *Darker* colors represent higher carbon biomass. Units are grams of carbon/30 m²

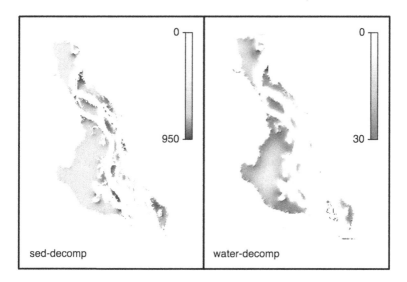

Fig. 9.7 NetLogo grayscale map output of decomposer stocks after a 57-day model simulation. *Darker* colors represent higher carbon biomass. Units are grams of carbon/30 m²

shallow, low-velocity backwater area to the left of the main channel (Fig. 9.4), locally known as Weaver Bottoms. By contrast, macrophytes exhibited more carbon biomass in deeper areas with higher water velocities (Fig. 9.4). The only exception to this pattern was found in the main channel, where velocities exceeded 1 m/s and complete scouring of macrophytes occurred.

9.4.2.2 Heterotrophic Consumer Stocks

Like primary producers, the consumers included in the model were present through-out the pool, with different subsets of consumers occupying different areas. Herbivore distribution mirrored that of phytoplankton (Fig. 9.5). A small amount of herbivores were present throughout the pool, but they gained the most biomass in shallow, low-velocity areas; additionally, water column consumer biomass distribu-tion matched the herbivore distribution (Fig. 9.5). Although their biomass was reduced in comparison with low-flow areas, water column consumer biomass was higher than herbivore biomass in deeper, high-velocity areas near the main channel. Finally, sediment consumer biomass distribution was similar to water column consumer biomass distribution (Fig. 9.5). However, sediment decomposers exhibited a greater decrease in biomass from shallow, low-velocity areas to deeper, high-velocity areas when compared with water decomposers.

9.4.2.3 Detritus, POC, and DOC

Detritus, POC, and DOC distributions throughout Pool 5 were similar to each other. All three carbon stocks exhibited the highest biomass in areas of moderate depth and velocity adjacent to the main channel (Fig. 9.6). None of these stocks accumu-lated biomass in the main river channel, and POC and DOC biomass were also very low in areas of low depth and velocity.

9.4.2.4 Decomposers

Decomposers in both the water column and sediments appeared to gain the most carbon biomass in areas of low to moderate depth and velocity in Pool 5 (Fig. 9.7). However, sediment decomposers appeared to inhabit more areas of the pool than water decomposers, and water decomposers exhibited increased biomass in areas adjacent to the main channel and to the river bank.

9.4.3 Validation

Preliminary model validation was performed using descriptive procedures. First, we ensured that the model accurately captures the key aspects of carbon flow in the Mississippi River and that the model was internally consistent and logical. No known aspects of the carbon cycle were violated in this model. The amount of carbon entering and exiting the model was conserved, and the model was free of programming errors.

To quantitatively validate this model, simulation results must be compared with field data. To accomplish this, spatial and temporal patterns from carbon distribution maps for each stock within Pool 5 must be compared with empirical data from Pool 5. Specifically, quantitative tests can be used to verify that spatial patterns in modeled stocks correlate with empirically observed spatial patterns in the same stock (e.g., vegetation bed size, configuration, total area, etc.). However, because data are still being compiled from the Long-Term Resource Monitoring Program of the US Geological Survey, only descriptive validation has been possible to date. Validation using more rigorous tests will continue as empirical data become available.

9.5   Discussion

Using existing models and data, we were able to build a preliminary spatially explicit carbon-based model of productivity for Pool 5 of the Upper Mississippi River. Our model suggests that the hydrology of the Mississippi River directly affects its food web. Simulation results indicate that the effect of depth on light availability was important to the success of primary producers in this simulation as they rely on light for carbon fixation. Furthermore, because macrophytes rely on their root system to anchor them and obtain nutrients, they were negatively affected by higher water velocities. The model dictated that above velocities of 1 m/s, macrophytes were completely scoured, inhibiting their survival regardless of light conditions. However, in shallow areas that had the lowest velocities and highest light availability, phytoplankton appeared to outcompete macrophytes. Therefore, macrophyte distribution in Pool 5 appears to be limited by biotic factors in addition to abiotic factors.

The data obtained from this simulation provide basic support for the RPM. Because the RPM states that consumers rely on autochthonous producers as a carbon resource, we would expect consumer carbon acquisition to depend heavily on producer carbon acquisition within this system. The model validates that expectation since consumers were able to acquire carbon in any location in the river that contained producers, and consumers did not run out of prey in 57 days under a relatively low river discharge of 2,097 m³/s. Additionally, the fact that outfluxes of carbon at the downstream end of Pool 5 were equal in size to carbon influxes at the upstream end suggests that Pool 5 did not function as a carbon sink for upstream carbon influxes or as a carbon source for downstream food webs. However, it is important to note that the model does not include allochthonous carbon inputs or simulate flood-pulse dynamics to directly test either the RCC or the FCC. Future iterations of this model must include these processes in order to more thoroughly explore their influences on river productivity.

This modeling approach is powerful and cost-effective because it capitalizes on hydraulic models that engineers are likely to employ in river development and restoration projects, previously collected ecological data, and an open source software modeling environment that is widely available at no cost. Details of the model processes can easily be modified depending on the user's needs as NetLogo runs on any reasonably current desktop computer under multiple operating systems.

Also, by using different hydraulic input maps, users can readily simulate different systems, and simulated time periods may be shortened or lengthened depending on the time step specified. Finally, this simple model can provide inputs to more complex individual-based or population-based simulation models.

Despite the many advantages of a simplified model, some limitations must be considered when evaluating this model's accuracy. For example, in our simulation of carbon flow, the hydraulic model does not account for waves and currents that are generated by outside forces such as wind or recreational boats (Bhowmik et al. 1982; Sparks 1984). Carbon removal from the river by external consumers such as fishermen, bald eagles, and other raptors, or by processes such as emergent insects being blown off-river, is not included in the model, and neither is the exchange of carbon dioxide between the river and the atmosphere (Rasera et al. 2008). Although the absence of details such as these may not dramatically affect the broad patterns simulated by this model, users may wish to include them in future versions of the model to add realism to the simulation.

Improved and validated versions of this spatially explicit river productivity model may be useful for predicting the impacts of major natural disturbances such as droughts and floods, and also human environmental interventions such as dam and levee construction. Such information could be crucial in mitigation of or preparation for the impacts of such disturbances. The impacts of restoration measures, such as reconnection of rivers and their floodplains or naturalization of water regimes, also could be evaluated using this type of model by making it possible to analyze the implications of alternate restoration scenarios. Simulation results may also help environmental agencies to justify more detailed modeling or data gathering prior to river development. The flexibility of our modeling approach allows application to a wide variety of systems and questions.

9.6 Conclusions

Mechanisms behind productivity in large river ecosystems are difficult to identify due to the large degree of spatial and temporal variability involved. Three different theories attempt to explain productivity in large rivers, but none describes a generalized mechanism that can be applied to all rivers and all seasons. Using NetLogo, we have developed a spatially explicit carbon-cycle model that simulates patterns of productivity in Pool 5 of the Mississippi River by incorporating both ecological and hydraulic components of the river. With further development and validation against empirical data, the model will be a simple and effective analytical tool for researchers studying a variety of river ecosystems.

Acknowledgments This is no. 37 in the contribution series of the National Great Rivers Research & Education Center.

References

Akanabi AA, Lian Y, Soong TW (1999) An analysis on managed flood storage options for selected levees along the lower Illinois River for enhancing flood protection. Report No. 4: flood Storage reservoirs and flooding on the Lower Illinois River, Illinois State Water Survey Contract Report 645, Illinois State Water Survey, Champaign

Allan JD, Castillo MM (2007) Stream ecology: structure and function of running waters. Springer, Dordrecht

Best EPH, Boyd WA (2008) A carbon flow-based modelling approach to ecophysiological processes and biomass dynamics of vallisneria americana, with applications to temperate and tropical water bodies. Ecol Model 217:117–131

Best EPH, Buzzelli CP, Bartell SM, Wetzel RL, Boyd WA, Doyle RD, Campbell KR (2001) Modeling submersed macrophyte growth in relation to underwater light climate: modeling approaches and application potential. Hydrobiologia 444:43–70

Bhowmik NG, Demissie M, Guo C-Y (1982) Waves generated by river traffic and wind on the Illinois and Mississippi Rivers. Water Resources Research Center Report UILU-WRC-82-0617, University of Illinois at Urbana-Champaign

Chambers PA, Prepas EE, Hamilton HR, Bothwell ML (1991) Velocity and its effects on aquatic macrophytes in flowing waters. Ecol Appl 1(3):249–257

Delaney RL, Craig MR (1997) Longitudinal changes in Mississippi River floodplain structure. U.S. Geological Survey Project Status Report no. PSR 97-02. U.S. Geological Survey, Upper Midwest Environmental Science Center, Onalaska

Dettmers JM, Wahl DH, Soluk DA, Gutreuter S (2001) Life in the fast lane: fish and foodweb structure in the main channel of large rivers. J N Am Benthol Soc 30(2):255–265

Doi H (2009) Spatial patterns of autochthonous and allochthonous resources in aquatic food webs. Popul Ecol 51(1):57–64

Francoeur SN (2001) Meta-analysis of lotic nutrient amendment experiments: detecting and quantifying subtle responses. J N Am Benthol Soc 20:358–368

Garbey C, Garbey M, Muller S (2006) Using modeling to improve models. Ecol Model 197:303–319

Gutreuter S, Bartels AM, Irons K, Sandheinrich SB (1999) Evaluation of the flood-pulse concept based on statistical models of growth of selected fishes of the Upper Mississippi River system. Can J Fish Aquat Sci 56(12):2282–2291

Herb WR, Stefan HG (2003) Integral growth of submersed macrophytes in varying light regimes. Ecol Model 168:77–100

Hoeinghaus DJ, Winemiller KO, Agostinho AA (2007) Landscape-scale hydrologic characteristics differentiate patterns of carbon flow in large-river food webs. Ecosystems 10(6):1019–1033

Houser JN, Richardson WB (2010) Nitrogen and phosphorous in the Upper Mississippi River: transport, processing, and effects on the river ecosystem. Hydrobiologia 640(1):71–88

Houser JN, Bierman DW, Burdis RM, Soeken-Gittinger LA (2010) Longitudinal trends and discontinuities in nutrients, chlorophyll, and suspended solids in the Upper Mississippi River: implications for transport, processing, and export by large rivers. Hydrobiologia 651(1):127–144

Huff DR (1986) Phytoplankton communities in navigation pool no. 7 of the upper Mississippi River. Hydrobiologia 136:47–56

Huisman J, Weissing FJ (1994) Light-limited growth and competition for light in well-mixed aquatic environments: an elementary model. Ecology 75(2):507–520

Junk WJ, Bayley PB, Sparks RE (1989) The flood pulse concept in river-floodplain systems. In: Dodge DP (ed) Proceedings of the international large river symposium (LARS), Canadian Special Publication of Fisheries and Aquatic Sciences, vol 106. pp 112–127

Kohler J (1995) Growth, production and losses of phytoplankton in the lowland River Spree: carbon balance. Freshwater Biol 34(3):501–512

Oliver RL, Merrick CJ (2006) Partitioning of river metabolism identifies phytoplankton as major contributor in the regulated Murray River (Australia). Freshwater Biol 51(6):1131–1148

Owens JL, Crumpton WG (1995) Primary production and light dynamics in an Upper Mississippi River backwater. Regul Rivers Res and Manage 11(2):185–192

Pace ML (1984) Zooplankton community structure, but not biomass, influence the phosphorous-chlorophyll a relationship. Can J Fish Aquat Sci 41:1089–1096

Park SS, Na Y, Uchrin CG (2003) An oxygen equivalent model for water quality dynamics in a macrophyte dominated river. Ecol Model 168:1–12

Power ME (2006) Environmental controls on food web regimes: a fluvial perspective. Prog Oceanogr 68(2–4):125–133

Power ME, Dietrich WE (2002) Food webs in river networks. Ecol Res 17:451–471

Rasera MDFL, Ballester MVR, Krusche AV, Salimon C, Montebelo LA, Alin SR, Victoria RL, Richey JE (2008) Small rivers in the southwestern Amazon and their role in CO_2 outgassing. Earth Interact 12(6)

Sand-Jensen K, Binzer T, Middelboe AL (2007) Scaling of photosynthetic production of aquatic macrophytes—a review. Oikos 116:280–294

Schramm HL, Eggleton MA (2006) Applicability of the flood-pulse concept in a temperate flood-plain river ecosystem: thermal and temporal components. River Res Appl 22(5):543–553

Sheldon F, Thoms MC (2006) In-channel geomorphic complexity: the key to the dynamics of organic matter in large dryland rivers? Geomorphology 77(3–4):270–285

Shih J, Alexander RB, Smith RA, Boyer EW, Schwarz GE, Chung S (2010) In-stream controls on total organic carbon in streams of the conterminous United States: an initial SPARROW model of land use and in-stream controls on total organic carbon in streams of the conterminous United States. U.S. Geological Survey Open-File Report 2010-1276, U.S. Geological Survey, Reston, Virginia

Sparks RE (1984) The role of contaminants in the decline of the Illinois River: implications for the Upper Mississippi. In: Weiner JG, Anderson RV, McConville DR (eds) Contaminants in the Upper Mississippi River. Butterworth Publishers, Stoneham, pp 25–66

Sparks RE (1985) Illinois large rivers NSF LTER progress report. Illinois Natural History Survey, Aquatic Biology Technical Report 1985(6). Illinois Natural History Survey, Champaign

Sparks RE, Bayley PB, Kohler SL, Osborne LL (1990) Disturbance and recovery of large flood-plain rivers. Environ Manage 14(5):699–709

Thorp JH, Delong MD (1994) The riverine productivity model: an heuristic view of carbon sources and organic processing in large rivers. Oikos 70:305–308

Tyser RW, Rogers SJ, Owens TW, Robinson LR (2001) Changes in backwater plant communities from 1975 to 1995 in Navigation Pool 8, Upper Mississippi River. Regul Rivers Res Manage 17(2):117–129

U.S. Army Corps of Engineers (2008) Surface Water Modeling System 10.0. U.S. Army Corps of Engineers, Coastal and Hydraulics Laboratory, Vicksburg. http://chl.erdc.usace.army.mil/sms

Vannote RL, Minshall GW, Cummins KW, Sedell JR, Cushing CE (1980) The river continuum concept. Can J Fish Aquat Sci 37:130–137

Vorosmarty CJ, McIntyre PB, Gessner MO, Dudgeon D, Prusevich A, Green P, Glidden S, Bunn SE, Sullivan CA, Reidy C, Liermann CR, Davies PM (2010) Global threats to human water security and river biodiversity. Nature 467:555–561

Wetzel RG (1984) Detrital dissolved and particulate organic carbon functions in aquatic ecosystems. Bull Mar Sci 35:503–509

Wetzel RG (2001) Limnology: lake and river ecosystems. Academic, Orlando

Wiegert RG (1975) Simulation models of ecosystems. Annu Rev Ecol Syst 6:611–638

Wilensky U (1999) NetLogo. Computer software. Center for Connected Learning and Computer-Based Modeling, Northwestern University, Evanston. http://ccl.northwestern.edu/netlogo/

Chapter 10
Simulating Gopher Tortoise Populations in Fragmented Landscapes: An Application of the FRAGGLE Model

Todd BenDor, James D. Westervelt, J.P. Aurambout, and William Meyer

10.1 Background

Habitat loss and associated habitat fragmentation is a growing problem worldwide and is considered to be one of the greatest threats to biodiversity as well as a primary cause of the current high rate of species extinction (Wu et al. 2003). Habitat fragmentation is defined as the process through which a natural habitat becomes divided into isolated small patches of complex geometrical form, within a sea of generally inhospitable land uses (Bunnell 1999; McComb 1999).

Fragmentation negatively affects many species, since smaller habitat patches support disproportionately smaller populations that are more vulnerable to local extinction than larger ones (Berec 2002; Hanski 1997; McComb 1999). By increasing patch isolation, fragmentation effectively decreases the gene pool of local populations, thereby favoring inbreeding depression and making the population more prone to demographic stochasticity (Berec 2002). It also modifies the quality of the remaining habitat by increasing edges and reducing core habitat, which can further

T. BenDor (✉)
Department of City and Regional Planning, University of North Carolina at Chapel Hill,
Campus Box 3140 New East Building, Chapel Hill, NC 27599-3140, USA
e-mail: bendor@unc.edu

J.D. Westervelt
Construction Engineering Research Laboratory, US Army Engineer Research
and Development Center, Champaign, IL, USA
e-mail: james.d.westervelt@usace.army.mil

J.P. Aurambout
Australian Department of Primary Industries, DPI Parkville Centre,
32 Lincoln Square North, Carlton 3053, Australia

W. Meyer
U.S. Army Engineer Research and Development Center, Construction Engineering
Research Laboratory, 2902 Newmark Drive, Champaign, IL 61822, USA

J.D. Westervelt and G.L. Cohen (eds.), *Ecologist-Developed Spatially Explicit*   171
Dynamic Landscape Models, Modeling Dynamic Systems,
DOI 10.1007/978-1-4614-1257-1_10, © Springer Science+Business Media, LLC 2012

increase biodiversity loss among core-dwelling species. Although fragmentation is strongly connected to habitat loss, this work focused on the connectivity issues that result from fragmentation.

10.1.1   The Role of Land-Use Change

Habitat loss and fragmentation result principally from the expansion of urban and agricultural land uses (Tigas et al. 2002). Land-use policies drastically and permanently affect animal and plant populations by modifying the landscapes in which they evolved. As a result, road construction, suburban development, and agricultural growth can alter species movement and dispersal patterns, potentially isolating and extirpating local populations (Bunnell 1999; Hanski 1997). Appropriate tools are needed to help planners consider development alternatives leading to minimal impacts on the environment, but the response of animal and plant species to land-use change (LUC) is highly complex, which complicates the development of such tools. Decisions based on oversimplification of species' needs or responses may lead to irreversible losses of biodiversity. Managers and planners increasingly seek decision-support tools based on simulation modeling to help assess the impacts of development alternatives (Jorgensen et al. 1996; van Daalen et al. 2002). The simulation modeling efforts reported here address the challenge of identifying the relative value of individual subpopulations with respect to their contribution to the diversity of future populations, given their response to proposed regional land management plans.

10.1.2   Gopher Tortoise Natural History

The native range of the gopher tortoise (*Gopherus polyphemus*) is found in parts of six southeastern states (Fig. 10.1). Today, although large populations still persist in Mississippi, Alabama, Georgia, and Florida, tortoise populations are declining throughout the species' range. Auffenberg and Franz (1982) estimated that in the last century gopher tortoise populations have declined by 80%. Additional declines of individual tortoise populations are occurring throughout the southeastern United States due to habitat destruction, fragmentation, and fire suppression (Lohoefener and Lohmeier 1984). This significant decline has led the U. S. Fish and Wildlife Service (FWS) to list the tortoise as a threatened species under the US Endangered Species Act of 1973 (ESA; 16 U.S.C. 35) in the western portion of its range in Louisiana, Mississippi, and Alabama (*Federal Register*, July 7, 1987). In the remainder of its range, it is considered by the Department of Defense to be a species at risk (SAR) and has been recommended for listing as threatened under the Endangered Species Act (Save Our Big Scrub Inc and Wild South 2006).

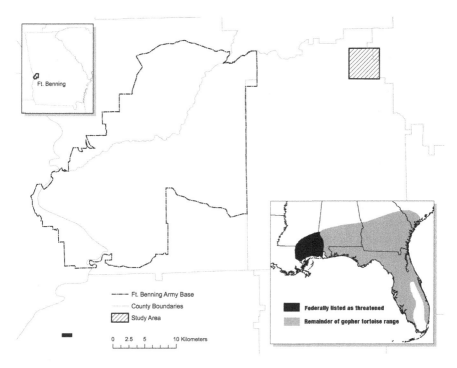

Fig. 10.1 Study area adjacent to Fort Benning with current gopher tortoise range and distribution

10.1.3 Habitat and Home Range Characteristics

Vegetation and soil associations that the gopher tortoise has been reported to colonize include sandhill, oak scrub, southeastern coastal plain, longleaf pine-oak (*Pinus palustris and Quercus* spp.), xeric hammock, pine flatwoods, dry prairies, mixed hardwood-pine communities, and ruderal areas (roadsides, fencerows, utility rights-of-way, pasture edges, clearings and fallow fields) (Auffenberg and Franz 1982; Burke 1989; Diemer 1986). Three conditions for survival must be met: well-drained sandy soils in which the tortoises can burrow, intermittent sunny areas in which they can bask and nest, and adequate low-growing forage. Fence rows, orchard edges, golf course roughs and edges, and some pastures may contain very dense populations (Auffenberg and Franz 1982). Home range sizes have been measured at between 0.008 and 9.167 ha (Mitchell 2005), with averages (in good habitat) reported to be 0.31 ha for adult females, 0.88 ha for adult males, and 0.05 and 0.01 ha for juveniles (Diemer 1992).

10.1.4   The Georgia Study Area

The study area for this project is east of the US Army's Fort Benning military installation, which is located in west-central Georgia, south of the City of Columbus (see Fig. 10.1). The study area occupies approximately 2,260 ha within Georgia, as shown by the hatched box in Fig. 10.1. This area could be considered for future gopher tortoise habitat protection since the predominant vegetation and soil associations consist of ideal tortoise habitat, particularly Sandhills (U.S. Geological Survey 2003a).

The study area has historically experienced fairly slow rates of urban LUC, and there is currently no evidence that this trend will change in the near- or mid-future. However, past agricultural land uses, including cotton farming and timber harvesting, have significantly fragmented the landscape. Land cover is shown in the base case scenario in Fig. 10.2a. FRAGGLE is one model that could help test the long-term implication of alternative land management strategies to protect populations of species-at-risk.

10.1.5   Issues in Fragmentation Modeling

A significant number of models have been developed for understanding the process of fragmentation and its effects. Although several species-specific modeling approaches have recently been used to determine potential species habitat within landscapes (Aurambout et al. 2005; Iverson et al. 1999; Mladenhoff et al. 1995; Ortega-Huerta and Peterson 2004), models that focus exclusively on habitat and species location can fail to account for demographic processes such as reproduction, mortality, immigration and emigration, rescue effects, and inbreeding depression (Hanski 1991). These processes can trigger potential time lags in habitat use patterns and can have extensive, long-term repercussions on the persistence of species within landscapes. Furthermore, habitat models do not often incorporate life stage analysis (Hastings 1996) or key landscape features selected by the species (McComb 1999). Consequently, the use of population models as complements to habitat models could improve predictions of species-specific demographic responses to environmental changes.

Although much of the research on species movement in fragmented landscapes has focused on corridors, less attention has been placed on studying species movements through adjacent matrices of habitat patches (With 1999). Moreover, as the amount of continuous natural habitat decreases below 60–80% of the landscape, connectivity between the remaining habitat patches becomes increasingly important for many species (McComb 1999). In order to evaluate trans-matrix species movements, the shortest distance between suitable habitat patches may not always provide the most appropriate estimate of movement behavior. Rather, an estimate of the permeability of individual land uses to the movement of species based on their specific biology may be necessary (Hanski 1997). Consequently, the creation of corridors may not be the only way to facilitate population movements between habitat patches.

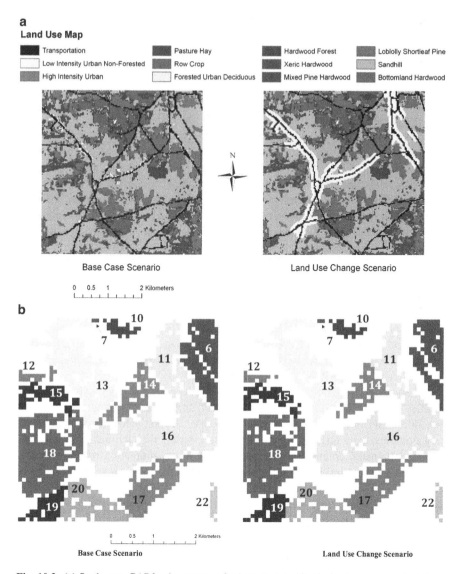

Fig. 10.2 (**a**) Study area GAP land cover map for base case and land-use change scenario. Minor land-use classes have been grouped together. Under the land-use change scenario, urban growth has occurred in corridors along major roads in the study area. These new areas are simulated to be impassable by the tortoise due to fencing constructed along the roads. Note that the Sandhill cells are core habitat areas. (**b**) Subpopulation areas maps. These areas are user-defined

10.1.6   Fragmentation Modeling Techniques

Researchers have identified two major classes of modeling techniques for habitat fragmentation: *agent-based models* (ABMs) and *cellular automata* (CA) models. ABMs, also referred to as individual-based models, have been effectively used to simulate the behavior of individual animals in the landscape (Grimm and Railsback 2005). They allow individuals to migrate and reproduce across an array of raster GIS habitat patches, while tracking the resulting diffusion of each individual's genetic characteristics. Individuals in ABMs move discretely, often with a stochastic component, in a chosen direction. An ABM can provide a high level of realism that makes the model easier for the user to grasp intuitively, but it needs high-performance computer technology to run the many simulations required to understand the effects of stochasticity on the probabilities of alternative population distributions. The Hargrove/Hoffman PATH ABM (Hargrove et al. 2001), for example, allows agents to migrate beyond the edges of suitable habitat into the interstitial landscape in search of new suitable habitat. Successful "walkers" look back on their paths and inform associated cells of the success. However, PATH is implemented on a supercomputer to accommodate the great number of simulations needed before results become meaningful. We previously attempted to apply a similar modeling approach using desktop computer technology, but simulation times were unacceptably long.

Alternatively, CA (Wolfram 1984) and other grid-cell modeling methods have been shown to be a viable alternative to the use of partial differential equations (a common mathematical technique) for modeling populations within landscapes (Darwen and Green 1996; Hanski and Ovaskainen 2000). CA have the capacity to handle complex boundaries and are effective for modeling dynamic processes (epidemics, biological invasions) in two-dimensional environments (Cannas et al. 2003; Karafyllidis 1998; Sirakoulis et al. 2000).

For our work, the primary advantage of the CA approach is that movement probabilities could be calculated and used directly. Instead of selecting a movement direction for an individual in ABM model, using a CA approach, a virtual individual may simultaneously migrate to more than one location, thereby making it possible to compute spatial connectivity measurements in one simulation run rather than hundreds. This capability provides a sort of shortcut for simulating stochasticity in movements, improving our understanding of how likely movement is to any specific habitat area. The average location of the individual becomes a probability surface, thereby negating the need for multiple model runs that may be required by an equivalent ABM (e.g., Monte Carlo analysis).

10.2 Objective

The objective of this research was to better understand the link between habitat fragmentation and subpopulation mixing through development and application of a spatial-dynamic model that can readily be adjusted and parameterized to capture the specific life-history and landscape characteristics associated with a variety of species and geographic areas. To verify the utility of the model, we studied the potential impact of a hypothetical LUC scenario in an area near Fort Benning, Georgia, on gopher tortoise (*G. polyphemus*) movements and populations.

10.3 Model Description

There is a wide variety of population viability analysis software available, including Vortex, RAMAS, Meta-X, SpaSIM, and ALEX (Brook et al. 1999, 2000; Lindenmayer et al. 1995; Spatialworks 2004), but their ready applicability to regional development and population fragmentation issues is uncertain. For purposes of development expediency, we constructed FRAGGLE using NetLogo 4.1 (Wilensky 1999).[1] Our approach incorporates a dynamic population model driven by a set of nonlinear difference equations (Deaton 2000; Ford 1999) into a grid cell structure in order to estimate strength of connections among gopher tortoise populations within a specified landscape. This method of representing local spatial interactions allows each grid cell to interact dynamically with its neighbors, thereby forming a dynamic, raster-based, spatially explicit simulation model. In order to parameterize each of the cellular models, several raster GIS input grid maps derived from land-use data are required. These cellular models are then run in parallel to simulate gopher tortoise population dynamics and movement of individuals within each habitat patch occurring in the landscape. By running the model on grid maps representing projected future land uses, the model provides indications of the potential long- and short-term responses of tortoise populations to regional human landscape development.

In FRAGGLE, a proportion of the population in each cell moves into adjoining patches based on factors including cell connectivity, relative crowding, and immigration rates. A population in each patch is defined by both the number of individuals in the cell and a vector representing the proportion of that population derived from individuals (or descended from individuals) from the original habitats. Population movement patterns are completely deterministic; they generate identical results for identical initial conditions, removing any need for multiple runs after the model is parameterized and calibrated. Populations in each patch are organized as nonmoving hatchlings, moving (but not reproducing) juveniles, and moving adults. Each life stage is parameterized to capture published characteristics of the target species.

10.3.1 Purpose

The purpose of FRAGGLE is to forecast the contribution of starting populations associated with patches (grouped into unique habitats) to the future populations in those patches. The desired end result is a set of values that indicate the relative contribution of the population in each patch to a future population in each patch.

---

[1] An operational copy of this model is available through http://extras.springer.com.

10.3.2   State Variables and Scales

10.3.2.1   Spatial Dynamic Framework

Using FRAGGLE, we simulate tortoise demographic and movement processes over a grid composed of regular 90-m square cells. This resolution (cell size of 0.81 ha) was chosen to approximate reasonable home range size for adults (1 ha). This size also allows for a straightforward combination of 30-m square GIS raster data, which is a common size for satellite imagery used in natural resource management.

10.3.2.2   Spatial Relationships and Input Data

In order to account for species-specific habitat considerations in estimating populations and their movement patterns in the landscape, the model requires four grid map inputs with identical geographical extent and cell size: (1) a habitat map, (2) a dispersal attractiveness map, (3) a mortality probability map, and (4) a subpopulation areas map. The habitat map identifies Sandhill as habitat, which is divided into core and edge habitat and initialized with tortoises at the start of a simulation. Core areas are those that support populations, with edges being those areas that transition between core habitat and non-habitat areas (90 m or 1 cell thick). The dispersal attractiveness map (index 0–1) represents the attractiveness of each non-habitat area to migrating animals. Animals motivated to migrate from a cell disperse to neighboring cells based on the relative attractiveness of those cells. The mortality probability map provides a value (0–1) representing the percentage of individuals that are lost to predation in a year (including human and nonhuman predation). Finally, the subpopulations areas map gives user-chosen areas, typically contiguous, which represent locations of more closely similar animals (numbered 1 to *n*).

10.3.3   Process Overview and Scheduling

10.3.3.1   Gopher Tortoise Population Model

We articulate the FRAGGLE population submodel around the three gopher tortoise life stages: hatchlings, juveniles, and adults. The population submodel (Fig. 10.3) centers on state variables (Ford 1999; Hannon and Ruth 1997) that track changes in the number of individuals at each life stage in a given cell. The content of each state variable (rectangles) is modified through input and output flows (double arrows with valves), whose functions are controlled through converters (circles) accounting for species-specific life history traits (Fig. 10.3). Feedback within the model is depicted through information links (arrows). This representation remains generic enough that it corresponds to the three-stage population structure of many vertebrate species.

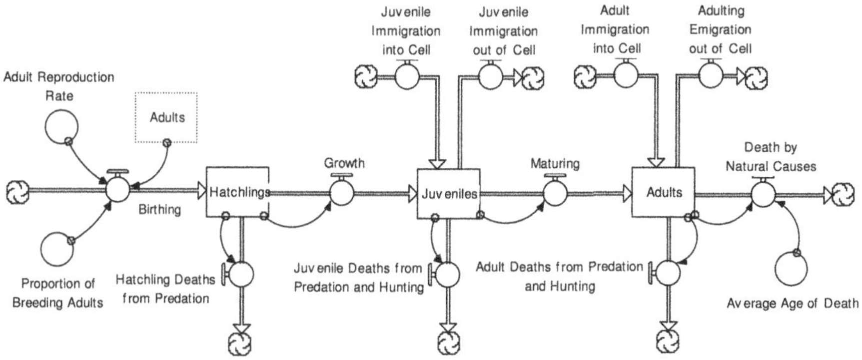

Fig. 10.3 STELLA diagram of the population submodel

Gopher tortoise hatchlings (under 1 year of age, given as *H*) represent a spatially static phase of the population model, since they rarely move beyond 8–15 m from their hatched location (McRae et al. 1981). They hatch in August and September, after the May and June breeding season (Iverson 1980; Landers et al. 1980; Mushinsky et al. 1997) as a result of mating by sexually mature adults (age over 20 years in many areas) (Landers et al. 1980), which only occurs in cells of suitable habitat, at a rate given in (10.1). We make no distinction between males and females as every individual able to secure territory is assumed to reproduce until death (tortoises maintain approximately 1:1 sex ratio and live 40–60 years, but can potentially exceed 150 years) (Landers et al. 1980). Assuming that tortoises mature to adults and suffer 3% mortality per year, then an adult will live on average to approximately 53 years.

$$H = \alpha\gamma \frac{A}{2} \tag{10.1}$$

Here, α represents the proportion of breeding adults, γ represents adult reproduction rate, and A represents the size of the adult population.

After 1 year, we assume that hatchlings turn into juveniles, and juveniles turn into breeding adults after a total of 20 years. Juveniles are assumed to have a fixed age class structure over their 19 years. Adult and juvenile tortoises represent mobile life stages during which movement between cells is possible. However, we assume adult tortoises tend to remain in their cell as long as it remains suitable for them, while a proportion of juveniles emigrate annually (see "Population Movement"). Once reaching adulthood, an individual remains in the adult stock until it moves out of the cell or dies. Here, we assume base reproduction rates are set as a random number bound between maximum and minimum values given as model inputs.

An important part of this model is the integration of separate death rates for each tortoise life stage. This is important to ensure that only surviving individuals can reproduce (Berec 2002). We varied death rates among hatchling (80% in core

Table 10.1 Habitat attractiveness and user-defined dispersal mortality rate by land cover

GAP 1 and class	Land-use category	Dispersal attractiveness (0–1)	Death rates during movement (0–1)
18	Transportation	0.180	0.09
22	Low-intensity urban non-forested	0.180	0.02
24	High-intensity urban	0.180	0.02
80	Pasture hay	0.593	0.02
83	Row crop	0.180	0.02
412	Hardwood forest	0.180	0.02
413	Xeric hardwood	0.263	0.02
432	Xeric mixed pine hardwood	0.537	0.02
434	Mixed pine hardwood	0.494	0.02
440	Loblolly shortleaf pine	0.764	0.02
512	Sandhill	1.000	0.02
900	Bottomland hardwood	0.312	0.02
930	Freshwater marsh	0.312	0.02

Our values are derived from Wright (1982)

habitat) (Landers et al. 1980; Wright 1982), juvenile (34%) (Wilson et al. 1994), and adult life stages to reflect their differential vulnerability to predators.

In contrast, mortality rates during dispersal are presumed to vary among specific land uses (Table 10.1). We simulate dispersion as a series of steps, across which the annual death rate due to predation is divided, as given in (10.2). Each dispersal step represents a type of sub-time step, whereby model calculations halt as adults and juveniles migrate. Here, adults are allowed to migrate for 16 steps, yielding a maximum potential dispersal distance of 1,440 m, while juveniles are allowed to migrate for 8 steps (720 m). After each dispersal step, a percent of the population in each cell is lost based on the cell dispersal mortality rates. After the dispersal steps complete during each time step (year), the model continues operation.

$$S = e^{\ln(1-r_a)/a_m} \qquad (10.2)$$

where S is the proportion of the population surviving, r_a is the adult death rate (3%), and a_m is the number of adult dispersal steps (16). This equation is replicated for juveniles using their respective death rates and dispersal steps. For example, distributing a 3% death rate over 20 dispersal steps results in a 99.848% survival rate between each dispersal step for adults. To end the annual cycle, all adult and juvenile populations in habitat are reduced to carrying capacity values and all remaining in non-habitat areas are presumed to die.

Additionally, various land uses present different environments and the tortoise death toll associated with crossing each cell depends on the associated land use/land cover. Therefore, we defined specific adult and juvenile dispersal death rates for each land use in Table 10.1. These dispersal death rates affect every migrating individual and are applied at every time step (1 year) during which adults and juveniles enter new cells.

10.3.4   Design Concepts

10.3.4.1   Habitat Carrying Capacity

The carrying capacity of an environment corresponds to the maximum number of individuals of a species that it can support for an indeterminate period of time. Species that are not limited by interactions with other species tend to increase their populations as far as their environment allows while remaining in a dynamic state of equilibrium and oscillating around the population carrying capacity value (Karafyllidis 1998). Given this, our model assumes that any cell of suitable habitat can provide a fixed amount of food and shelter for tortoise populations. The maximum individuals per cell (M) is calculated from model parameters as follows:

$$M = g h_s c_s \qquad (10.3)$$

where g is the number of animals per home range (3), c_s is the cell size (0.81 ha), and h_s is the size of the home range (1 ha).

   We assume a maximum carrying capacity for a suitable habitat cell of three adults due to territoriality effects (McRae et al. 1981), leading to a value of 2.43 for M.

10.3.4.2   Movement

The decision to move is triggered by a specified inherent likelihood for adults and juveniles to move in combination with an analysis of juvenile overcrowding in which a juvenile migrates if it exceeds the cell's carrying capacity. The decision on what direction to move can be completely random ("directed-random") or optimized. FRAGGLE allows for the fact that some species are able to sense distant habitat, and therefore take a more or less optimized path toward those areas. In the current example, however, the gopher tortoise moves in a "directed-random" manner since the probability of moving any direction is based on the relative dispersal attractiveness of adjacent habitat compared to the combined attractiveness of all adjacent habitat cells.

$$P_i = \frac{W_i}{\sum_{n-1}^{m} W_i} \qquad (10.4)$$

P_i: proportion of the cell population emigrating to a given neighbor
W_i: weight of the land use situated in the ith direction of the target cell
m: number of adjacent neighboring cells, assumed here to be eight

10.3.4.3   Mixing of Subpopulations

Consider a simple FRAGGLE model with two populations. For this model, each cell tracks a 2-element vector, which at model initialization is set to [1.0   0.0] for populations within Area 1 and [0.0   1.0] for populations within Area 2. If, in the course of model simulation, migrating populations can successfully cross the non-habitat space between the areas, these values will change to provide a running count of the percentage of contribution of the original designated areas to the population at every location.

Consider, for example, a five-area model. At the outset, cells with animals in Area 1 will be assigned the vector [1.0   0.0   0.0   0.0   0.0], which indicates that 100% of the lineage originated in Area 1, and 0% from the other 5 areas. After some years of simulation that vector might shift to [0.6   0.1   0.0   0.3   0.0] (60% from Area 1; 10% from Area 2; 0% from Areas 3 and 5; and 30% from Area 4) indicating a significant influx of population from Area 4, some influx from Area 2, and apparent isolation from Areas 3 and 5. This information along with population sizes can be used to identify the original areas that provided the richest lineage sources (and strongest movement capabilities) in the study area.

For example, consider a situation with four separate areas with populations of 10, 5, 8, and 20 individuals, respectively. The state of this original population can be defined as:

$$\begin{bmatrix} 1.0 & 0.0 & 0.0 & 0.0 \\ 0.0 & 1.0 & 0.0 & 0.0 \\ 0.0 & 0.0 & 1.0 & 0.0 \\ 0.0 & 0.0 & 0.0 & 1.0 \end{bmatrix} \bullet \begin{bmatrix} 10 \\ 5 \\ 8 \\ 20 \end{bmatrix} = \begin{bmatrix} 10 \\ 5 \\ 8 \\ 20 \end{bmatrix} \qquad (10.5)$$

By multiplying the population vector with the distribution, we get an array of values that we will call *individual-equivalents*. These indicate the total contribution of an original area to the future populations. For example, an Area 1 with an original population of 10 individuals may be responsible in the future for 80% of the lineage of 5 individuals in that same area, 30% of the lineage of 10 individuals in Area 2, and 20% of the lineage of 15 individuals in Area 3. This would calculate to 0.8 * 5+0.3 * 10+0.2 * 20=11.0 individuals, which suggests an overall increase in the population of the area that originated in Area 1, making Area 1 a population source. Of course, at the start of the simulation, these are identical to the population distribution. It indicates that there are 10 individual-equivalents of genes from Area 1, 5 from Area 2, 8 from Area 3, and 20 from Area 4. At the end of the simulation, the population (vertical vector) and lineage contribution in each area could be as follows:

$$\begin{bmatrix} 0.6 & 0.1 & 0.0 & 0.4 \\ 0.1 & 0.8 & 0.1 & 0.0 \\ 0.0 & 0.1 & 0.9 & 0.0 \\ 0.3 & 0.0 & 0.0 & 0.6 \end{bmatrix} \bullet \begin{bmatrix} 10 \\ 1 \\ 2 \\ 10 \end{bmatrix} = \begin{bmatrix} 10.1 \\ 2.0 \\ 1.9 \\ 9.0 \end{bmatrix} \qquad (10.6)$$

The columns in the matrix give the proportion of the lineage coming from each original area. For example, at the end of the simulation there are 10 individuals in

Area 1, and 60% of the lineage came from Area 1 at the start of the simulation, 10% from Area 2, and 30% from Area 4. Note that the columns each sum to 1. Multiplying these matrices yields the individual-equivalents from each of the original areas (10.1 from Area 1, 2.0 from Area 2, 1.9 from Area 3, and 9.0 from Area 4).

To understand the contribution of the original subpopulation areas to the final population, we divide the values in the final individual-equivalents array with the first and get:

$$[1.01 \quad 0.40 \quad 0.28 \quad 0.45] \tag{10.7}$$

The final vector suggests that original individuals in Area 1 were the best at contributing to the future population. Area 4 began with 20 individual-equivalents, but dropped 55% to only 9 individual-equivalents. This analytical technique addresses the basic question of which areas are able to best contribute to future generations on a given landscape.

10.3.4.4   Subpopulation Movement Across the Landscape

We assume that movement of a small percentage of tortoises occurs in underpopulated areas, and that a higher percentage moves when the population exceeds the cell carrying capacity. This density-dependent approach accommodates the forcing of juveniles and adults to search for new territories. In this model, juveniles compete among themselves for space (as do adults), but adults and juveniles do not compete with each other.

During tortoise movement, lineage must be traced to enable us to estimate the contribution of each original population through time. This is accomplished by continually merging incoming dispersing individuals with the current local population to update the overall lineage composition of a cell. As we saw above, lineages are tracked with lineage arrays associated with each cell. Each array contains a position to identify the lineage contribution of the original subpopulation areas.

Consider an example system with four subpopulations, where we aim to capture the movement of individuals in and out of cells. Assume there are 5.2 individuals in one cell with the lineage array [0.1 0.2 0.3 0.4]. Of those individuals, 2.3 will move into a neighboring cell that contains 2.7 individuals associated with the lineage array [0.2 0.3 0.5 0.0]. The state of the system after this move will find 2.9 individuals in the first cell with the identical original lineage array [0.1 0.2 0.3 0.4]. The second cell will have 5.0 individuals with a combined lineage array based on the proportion of original and incoming animals and their original arrays as follows:

$$\left(\frac{2.3}{5.0}\right) \bullet [0.1 \quad 0.2 \quad 0.3 \quad 0.4] + \left(\frac{2.7}{5.0}\right) \bullet [0.2 \quad 0.3 \quad 0.5 \quad 0.0]$$
$$= [0.154 \quad 0.255 \quad 0.408 \quad 0.184].$$

Note that the fractional vector coefficients in the final calculation sum to 1.0, representing a full accounting of the source of 100% of the lineage from the original lineage areas. A similar type of calculation is also used to track the average age of animals in each cell.

Consider another simple example of two circular subpopulation areas arranged east–west next to each other. Adults are randomly assigned to each cell within each of these areas and are assigned lineage markers [1 0] in the left area and [0 1] in the right. At step 1, the population in the cells associated with the left area originated 100% from the left area (represented by the solid black) and the populations in the cells associated with the right area are white, representing 100% origination from the right area. By step 20, through dispersal simulation, dispersing individuals from both areas have begun to move into the other area, which is indicated by the lightening of the cells in the left area and the darkening of the cells in the right area. The rate of lineage mixing between the two core habitat areas emerges from the model based on the starting number of individuals in the original habitats, the distance between habitats, and the differential ability of animals to cross each of the intervening cells. A separate population contribution array and average age is maintained for hatchlings, juveniles, and adults, but movements are only calculated for mobile juveniles and adults.

To measure the extent of population mixing within individual cells and across original lineage areas, we created the lineage mixing index. The value can range from 0, indicating that there is no lineage sharing with other areas, to 1, indicating an equal mixing with all other areas. It is calculated as follows:

$$\sum_{x=1}^{n} \min\left[c_x * \frac{T/n}{T}, \frac{T-c_x}{T-T/n} \right] \tag{10.8}$$

where:

T = Total population in the subpopulation area
c_x = The total contribution of the original subpopulation in area x to this area
n = The number of subpopulation areas

When the total population originates from a single area, this results in an index of 0. Conversely, when that total population is equally derived from all other areas, the index will be 1.0. This calculation can be used for any spatial unit; in this analysis it is used for both cells and the original areas.

10.3.5 Initialization

To start a simulation, populations of gopher tortoises at each life stage are initialized within each cell identified by the habitat as occupied. A reasonable January population structure is initialized in each cell (including 0–6 hatchlings, 1–4 juveniles, and

Table 10.2 Model parameterization

Model parameter	Value	Unit	Citation/assumption
Proportion of adults breeding	100	Percent	Assumption (all are sexually active)
Hatchlings initialized in core cells	0	Count	Assumption
Hatchlings predation rate in edge/core	81	% Pop/year	Landers et al. (1980); Wright (1982)
Hatchlings predation rate outside habitat	100	% Pop/year	Landers et al. (1980); Wright (1982)
Juveniles initialized in core cells	5–10	Count	Assumption
Juvenile predation rate in edge/core	20/10	% Pop/year	Wilson (1991)
Juvenile predation rate outside habitat	100	% Pop/year	
Juvenile hunting rate	0	% Pop/year	
Juvenile capacity per home range	12		McRae et al. (1981)
Juvenile dispersal steps per year for those moving	8	90-m steps	Assumption
Adults initialized in core cells	0.1–0.3		Assumption
Adult predation rate in edge/core	3	% Pop/year	Taylor (1982); Lohoefener and Lohmeier (1984), assumption
Adult predation rate outside habitat	100	% Pop/year	
Adult hunting rate	0	% Pop/year	
Adult capacity per home range	3	Count	McRae et al. (1981)
Reproduction age	20	Years	Landers (1982)
Hatchlings per adult per year	2–4	Count	Iverson (1980); Smith (1995); Smith et al. (1997)
Animals per home range	3		McRae et al. (1981)
Home range size	1	Ha	Diemer (1992)
Cell size	0.81	Ha	U.S. Geological Survey (2003a, b)
Percentage of adults/juveniles that move each year from under-populated locations	5	% Pop/year	5% for distances greater than 250 m based on assumptions gained from reviewing (Auffenberg and Iverson 1979; McRae et al. 1981; Diemer 1986)
Adult dispersal steps per year for those moving	16	90-m steps	Based on an interpretation of McRae et al. (1981); Diemer (1992)

Estimates were either taken directly from cited sources or were estimated based on information in these sources

1–3 adults). Table 10.2 lists the FRAGGLE biological parameters that are set for each initialized tortoise.

Tortoise populations in each cell are initialized with a lineage array indicating that 100% of their lineages are associated with their starting area. The model is then run for a 100-year simulation.

10.3.6   Input

We begin with Gap Analysis Program (GAP) land use/cover maps (based on the National Land Class Dataset—NLCD) produced by the U.S. Geological Survey (2003a, b). These maps are derived from Landsat Thematic Mapper (TM) imagery with a spatial resolution of 30×30 m (later aggregated to 90×90 m using the nearest-neighbor and averaging algorithms, as appropriate). The Georgia GAP map was created in two stages, beginning with the production of an 18-class Anderson-level map (Anderson et al. 1976), which was then refined into a much more detailed 44-class land cover map. The overall state-wide accuracy of this two-step process is 75.46% (U.S. Geological Survey 2003a).

A survey of current literature was conducted to determine which of the 44 classes in the Georgia GAP map most closely resembled vegetation and soil associations consistent with gopher tortoise habitat. A dispersal attractiveness value was assigned to each land class in terms of its suitability as gopher tortoise dispersal corridors. The habitat map was generated by selecting Sandhill areas (land class 512) as suitable habitat (given a dispersal attractiveness value of 1.0, or 100%), while all other values were assigned as non-habitat (see Fig. 10.2). Dispersal attractiveness and mortality probability maps are developed as a cross-reference of information in Table 10.1, which characterizes the utility or attractiveness of each land-use category for supporting dispersal of the target species, as well as the death rate experienced while crossing unsuitable habitat. Our mortality rate information was derived from values given in Wright (1982) for tortoises 5 years and older.

Finally, we generated the subpopulation areas map (Fig. 10.2b) by clumping habitat areas into contiguous areas, with each area assigned a unique small integer as an identifier. The FRAGGLE model is designed to forecast how populations from each area are mixed into the future populations of all areas. Note that any user-chosen subdivision of the areas could be used. To this point, all GIS operations used a 30-m resolution. Simulation model input maps were resampled to a 90-m resolution using a nearest-neighbor approach. The habitat map and the subpopulation input map are shown in Fig. 10.2.

10.4   Simulation Experiments

10.4.1   Summary of Simulations

Simulations were run to analyze two subpopulation exchange scenarios under differing land-use configurations. The first scenario retains existing land use and land-cover patterns, while the second has undergone a plausible, but artificial, LUC. Habitat and subpopulation area input maps to the model are displayed in Fig. 10.2a, b. While the habitat loss associated with the LUC scenario does not affect many subpopulation areas, it reduces overall habitat by 10.2% and reduces the size of some lineage areas by as much as 18% (e.g., Areas 6, 11, and 14).

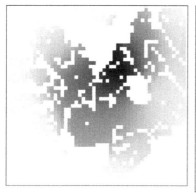

Base Case Scenario   Land Use Change Scenario

Fig. 10.4 Dispersal spread map for individuals originating in Area 14. *Darker* areas represent higher mixing through dispersal. The color scale ranges linearly from nearly one individual-equivalent (*black*) to zero individual-equivalents (*white*)

This model was developed to test the consequences of alternative regional planning scenarios on the interconnectivity of remaining areas of suitable habitat. Scenarios might reflect a county zoning that promotes the development of housing and commercial uses in the area. A LUC scenario was developed to test the ability of the model to illuminate consequences in the projected lineage exchanges among subpopulations. In this scenario, new development would be restricted to areas that are within 200 m of urban land uses and within 100 m of existing roads (Fig. 10.2a). The projected result of this scenario is that the attractiveness weight for tortoises through these areas drops to 0, creating a dispersal barrier.

10.4.2   Results

Figure 10.4 displays the extent of the movement of the Area 14 population marker across the landscape at the end of 100 years for each of the scenarios. Populations from each of the subpopulation areas moved into other areas and intersperse their subpopulations with other populations. The implications of the two scenarios with respect to the contributions of original areas to future populations are captured in Tables 10.3 and 10.4. Columns 1–3 of these tables show the original areas followed by the total initial and final populations in those areas. The fourth column provides the subpopulation mixing index at the end of the 100-year simulation. The remaining columns show the distribution of individuals within each area with lineages originating from the areas indicated by the column headings. For example, in the base case scenario, Area 6 started with 257 individuals and ended with 334, while lineages were mixed with individual-equivalents from 7 areas. While most of the lineages in Area 6 originated in Area 6 (222 individual-equivalents), 58 individual-equivalents emigrated from Area 11, 48 from Area 16, 5 from Area 14, and 1 from

Table 10.3 Dispersal behavior in base case scenario

Area	Start individuals	End individuals	Mixing index	Areas contributing to population lineages													
				6	7	10	11	12	13	14	15	16	17	18	19	20	22
6	257	334	0.186	**222**	**0**	0	**58**	**0**	**0**	**5**	0	**48**	**1**	**0**	**0**	**0**	**0**
7	5	5	0.024	0	**0**	2	0	0	3	0	0	0	0	0	0	0	0
10	52	45	0.069	1	1	**35**	3	0	5	0	0	0	0	0	0	0	0
11	258	244	0.247	87	0	1	**100**	0	1	24	0	30	1	0	0	0	0
12	35	35	0.124	0	0	0	0	**3**	14	0	14	0	0	3	0	0	0
13	516	510	0.366	0	1	2	1	5	**388**	29	26	40	1	16	0	0	0
14	187	157	0.304	7	0	0	24	0	25	**47**	1	49	2	2	0	0	0
15	131	130	0.238	0	0	0	0	7	40	3	**49**	5	0	26	0	0	0
16	852	788	0.504	47	0	0	19	0	27	37	2	**580**	55	16	1	6	0
17	290	253	0.191	2	0	0	1	0	1	3	0	107	**99**	1	1	39	0
18	547	521	0.420	0	0	0	0	1	25	5	16	32	1	**385**	34	24	0
19	119	119	0.161	0	0	0	0	0	0	0	0	1	0	53	**36**	28	0
20	260	256	0.285	0	0	0	0	0	0	0	0	9	26	43	36	**141**	0
22	30	30	0.010	0	0	0	0	0	0	0	0	1	0	0	0	0	**29**
Individual-equivalents				366	3	40	205	16	529	152	109	901	186	545	107	238	29
Percent change				43	-40	-5	-7	-44	24	-5	2	27	-21	23	10	11	-11

Bold values trace the row-column intersections associated with the same area. Values are all rounded and resulting zero-value entries are removed to improve readability. "Individual-equivalents" indicates the total number of individuals in the map from each subpopulation area after 100 years of spread and mating. "% change" indicates the total increase or decrease of population originating in each area. Large decreases and areas with no dispersal flux indicate areas in danger of genetic drift or depression

Table 10.4 Dispersal behavior in land-use change scenario

Area	Start individuals	End individuals	Mixing index	Areas contributing to population lineages													
				6	7	10	11	12	13	14	15	16	17	18	19	20	22
6	213	267	0.090	**247**	0	0	2	0	0	0	0	18	0	0	0	0	0
7	4	5	0.025	0	**0**	2	0	0	3	0	0	0	0	0	0	0	0
10	48	55	0.073	0	1	**44**	4	0	5	1	0	0	0	0	0	0	0
11	174	216	0.191	3	0	1	**143**	0	2	40	0	26	1	0	0	0	0
12	21	25	0.064	0	0	0	0	**4**	0	0	17	0	0	4	0	0	0
13	364	443	0.132	0	1	3	2	0	**407**	26	2	2	0	0	0	0	0
14	133	170	0.230	0	0	0	50	0	31	**75**	0	13	0	0	0	0	0
15	102	119	0.144	0	0	0	0	9	2	0	**71**	0	0	37	0	0	0
16	671	826	0.335	20	0	0	17	0	1	8	0	**703**	66	4	0	7	0
17	261	316	0.169	1	0	0	1	0	0	0	0	127	**147**	0	1	39	0
18	427	522	0.242	0	0	0	0	2	0	0	28	5	0	**449**	29	10	0
19	85	107	0.158	0	0	0	0	0	0	0	0	0	0	45	**37**	24	0
20	187	238	0.282	0	0	0	0	0	0	0	0	9	35	17	31	**146**	0
22	34	32	0.008	0	0	0	0	0	0	0	0	1	0	0	0	0	**31**
Individual-equivalents				271	3	50	219	14	452	151	118	904	250	556	98	225	31
Percent change				27	-34	2	26	-33	24	13	15	35	-4	30	16	20	-9

Area 17. The two bottom rows show the total and percentage change in the number of individual-equivalents originating from each of the original area populations during the course of the simulation. In Table 10.3 these values suggest that Areas 7 and 12 will not contribute much to the lineages of the region in the future. Conversely, Areas 6, 13, 16, and 18 appear to have the strongest connectivity to other areas. Area 22 is the most isolated, with virtually no input from (or output to) other populations. The mixing index, a measure of the level of connectedness between each area, is provided in column 4 in Tables 10.3 and 10.4.

Area 22 is associated with a mixing index of 0.010, which suggests the potential for genetic drift and inbreeding depression that may eventually result in increased death rates and reductions in fecundity (birth rates). Areas 7 and 10 are also relatively isolated, indicated by indices of 0.024 and 0.69, respectively. In this model, the largest spatial Areas (13, 16, and 18) are also sufficiently connected to nearby areas to enable them to be the greatest contributors and receivers of dispersing individuals from neighboring areas (indices of 0.366, 0.504, and 0.420, respectively). These large areas, and their most important neighbors (e.g., Areas 12, 14, 15, 17, 19, and 20), are perhaps the best to target for preservation to ensure a healthy population of tortoises far into the future.

Consider now the LUC scenario using Table 10.4 and the images in Fig. 10.4, which compare the spread of subpopulations in both scenarios. The figure shows a visibly significant difference in the distribution of subpopulations from Area 14 into Area 16 to the south and Area 13 to the west. At the end of a 100-year run, the total number of animals is 3,342, compared to 3,617 in the base case scenario, a drop of over 7%. Subpopulation mixing in 11 of the 14 areas dropped (only Areas 7 and 10 slightly increased), reflecting an increase in habitat fragmentation. This is reflected in an overall decrease in the average mixing index in the areas from 0.223 to 0.153, a drop of 31%. The subpopulation areas that become most isolated are Area 13 with a 64% drop in the mixing index, Area 6 with a 52% drop, Area 12 with a 49% drop, and Area 18 with a 42% drop.

10.5 Discussion

Fragmentation leads to the isolation of populations, which increases the likelihood of inbreeding, and eventually can contribute to significant reductions in population abundance, genetic diversity, and, potentially, local extirpation. The development of FRAGGLE was motivated by the need to rapidly predict the impact of proposed land management decisions on the habitat fragmentation of a target species.

FRAGGLE provides an excellent method for rapidly projecting the redistribution of subpopulations (and their progeny) within a metapopulation over time in response to alternative land management practices. A population is divided into subpopulations and each is provided a unique marker. Through dispersal of the animal as a juvenile and as an adult, those markers are inserted into neighboring populations. At any point in the simulation, the percentages of the population in each cell or area possessing the subpopulation markers from each of the original populations can be estimated. At any point in the simulation, each cell location contains the total number of individuals (a floating-point number) as hatchlings, juveniles, and adults, and the proportional contribution of the original populations to each life-stage. For example, at some point during simulation time, a cell might have 10.34 adults with a contribution ratio from all areas in the simulation (which sum to 100%). This information can be spatially aggregated as desired.

10.6 Conclusions

10.6.1 Utility for Other Species

FRAGGLE should be directly applicable for evaluating the importance of core habitat patches and dispersal paths for other species. As each species is potentially very different in its dispersal behavior, the model may need to be adapted. To facilitate adaptations, we developed the model within the NetLogo spatially explicit simulation modeling system, making it very accessible to a wide range of researchers. NetLogo is readily available for no cost, and provides model compilation, editing, and visualization capabilities, which allows the model developer to focus on the system being simulated, not software acquisition and training.

A number of model adaptations might be anticipated, depending on the species. For the gopher tortoise, FRAGGLE allows an animal to choose a dispersal path based only on the relative dispersal attractiveness of the current location and the eight immediate neighbors. Larger and more wide-ranging animals may need modifications that allow for a broader perception of the landscape. Additionally, various levels of learning could be supported for more cognitively advanced species.

An important consideration when attempting to ensure the ability of a natural population to persist is how to maintain a sufficient level of diversity. When combined with knowledge of the impact of inbreeding on fecundity and survival rates, the subpopulation mixing coefficients calculated in the model could potentially be used to predict population viability rates.

This model assumes land management activities to maintain the current system state, including the halting of natural succession. Gopher tortoises in and around the region of our study area require non-climax habitat that was naturally maintained through fire. While an overstory can be beneficial in moderating summer temperatures, a well-developed mid-story will eventually reduce the ground vegetation required as forage to the point that tortoises will abandon the site. In past centuries, natural and human-induced fire ensured an ever-changing mosaic of suitable habitat. With current fire suppression to protect human lives, homes, and agriculture, it becomes imperative to insert human-induced management into any future tortoise habitat. This model assumes that through targeted management habitat will remain constant. Hence, in this model, vegetation growth or succession, weather, and wildfire are not simulated. However, for other species, simulation of these or other landscape processes may be required.

10.6.2 Policy Implications

With the steadily increasing fragmentation of natural habitat within the matrix of human modified landscapes, it has become critical to understand a species' movement strategies and patterns in order to create successful species management regimes. Increasingly, planners and environmental managers need the ability to anticipate possible repercussions of changes in local and regional land-use policies on sensitive animal species.

By simulating gopher tortoise population dynamics and movement of individuals through the landscape, environmental managers can create land-use policies that maximize the likelihood of population persistence. Examples of these policies include the placement and design of corridors or road overpasses in order to allow the maximum number of individuals to cross inhospitable patches of land. Evidence has shown that landscape modifications, while minimal at the scale of a species' home range, can have dramatic consequences on individual movement and behavior (e.g., the passage of a highway in the middle of the grizzly bear home range) (Gibeau and Herrero 1998). However, fragmentation models must consider many factors, including the relative cost and effectiveness of corridors (Simberloff and Cox 1987). Furthermore, fragmentation models must provide managers with the ability to compare the relative ecological value of potential habitat restoration sites for particular species and then choose optimal locations.

10.6.3 Directions for Future Gopher Tortoise Research

The results of this model can be improved through further long-term studies of the gopher tortoise. Although it is a well-studied species, its longevity and generation time increase the difficulty in conducting population studies. This model requires information concerning dispersal distances during habitat searches, as well as a better understanding of how much information an animal can use or obtain to strategize and maximize its chances for dispersal success. The projections yielded by this model will be vastly improved if our associated assumptions were replaced with the results of future field research.

Acknowledgments The authors thank the Land-use Evolution and Impact Assessment Model (LEAM) group and the U.S Army Engineer Research and Development Center, Construction Engineering Research Laboratory (ERDC-CERL) for providing funding and data support and computing capacity. We also thank Jeff Terstriep and Yong Wook Kim for their help with computational issues.

References

Anderson JR, Hardy EE, Roach JT, Witmer RE (1976) A land use and land cover classification system for use with remote sensor data. Geological Survey Professional Paper 964

Auffenberg W, Franz R (1982) The status and distribution of the gopher tortoise (*Gopherus polyphemus*), North American tortoises: conservation and ecology. US Fish and Wildlife Service. Wildlife Research Report, No. 12

Auffenberg W, Iverson JB (1979) Demography of terrestrial turtles. In: Harless M, Norlock N (eds) Turtles: research and perspectives. Wiley-International, New York, p 718

Aurambout J, Endress A, Deal B (2005) A spatial model to estimate habitat fragmentation and its consequences on long-term persistence of animal populations. Environ Monit Assess 109(1):199

Berec L (2002) Techniques of spatially explicit individual-based models: construction, simulation, and mean-filed analysis. Ecol Model 150:55–81

Brook B, Cannon J, Lacy R, Mirande C, Frankham R (1999) Comparison of the population viability analysis packages GAPPS, INMAT, RAMAS and VORTEX for the whooping crane (*Grus americana*). Anim Conserv 2:23–31

Brook BW, Burgman MA, Frankham R (2000) Differences and congruencies between PVA packages: the importance of sex ratio for predictions of extinction risk. Conserv Ecol 4(1):6

Bunnell FL (1999) What habitat is an island? In: Rochelle JA, Lehman LA, Wisniewski J (eds) Forest wildlife and fragmentation management implications. Brill, Leiden, pp 1–31

Burke RL (1989) Florida gopher tortoise relocation: overview and case study. Biol Conserv 48:298–309

Cannas SA, Marco DE, Páez SA (2003) Modeling biological invasions: species traits, species interactions, and habitat heterogeneity. Math Biosci 183:93–110

Darwen PJ, Green DG (1996) Viability of populations in a landscape. Ecol Model 85:165–171

Deaton MJW (2000) Dynamic modeling of environmental systems. Springer, New York

Diemer JE (1986) The ecology and management of the gopher tortoise in the Southeastern United States. Herpetologica 42(1):125–133

Diemer JE (1992) Home range and movements of the tortoise (*Gopherus polyphemus*) in northern Florida. J Herpetol 26(2):158–165

Ford A (1999) Modeling the environment: an introduction to system dynamics modeling of environmental systems. Island Press, Washington

Gibeau ML, Herrero S (1998) Roads, rails and grizzly bears in the Bow River Valley, Alberta. In: Proceedings International Conference on Ecology and Transportation, ed. F. Department of Transportation. Tallahassee

Grimm V, Railsback SF (2005) Individual-based modeling and ecology. Princeton University Press, Princeton

Hannon B, Ruth M (1997) Modeling dynamic biological systems. Springer, New York, p 399

Hanski I (1991) Single-species metapopulation dynamics—concepts, models and observations. Biol J Linn Soc 42(1–2):17–38

Hanski I (1997) Predictive and practical metapopulation models: the incidence function approach. In: Tilman D, Kareiva E (eds) Spatial ecology the role of space in population dynamics and interspecific interactions. Princeton University Press, Princeton, pp 21–45

Hanski I, Ovaskainen O (2000) The metapopulation capacity of a fragmented landscape. Nature 404:755–758

Hargrove WW, Hoffman FM, Sterling TL (2001) The do-it-yourself supercomputer. Sci Am 256(2): 72–79

Hastings A (1996) Models of spatial spread: is the theory complete? Ecol Soc Am 77(6):1675–1679

Iverson JB (1980) The reproductive biology of *Gopherus polyphemus* (Chelonia: Testudinidae). Am Midl Nat 103(2):353–359

Iverson LR, Prasad A, Schwartz MW (1999) Modeling potential future individuals tree-species distributions in the eastern United States under a climate change scenario: case study with *Pinus virginiana*. Ecol Model 115:77–93

Jorgensen SE, Halling-Sorensen B, Nielsen SN (1996) Handbook of environmental and ecological modeling. CRC Press, Boca Raton

Karafyllidis I (1998) A model for the influence of the greenhouse effect on insect and microorganism geographical distribution and population dynamics. Biosystems 45:1–10

Landers JL (1982) Growth and maturity of gopher tortoises in southwestern Georgia. Bull Fla State Mus 27:81–110

Landers JL, Garner JA, McRae WA (1980) Reproduction of gopher tortoises (*Gopherus polyphemus*) in southwestern Georgia. Herpetologica 36(4):353–361

Lindenmayer DB, Burgman MA, Akçakaya HR, Lacy RC, Possingham HP (1995) A review of the generic computer programs ALEX, RAMAS/space and VORTEX for modelling the viability of wildlife metapopulations. Ecol Model 82(2):161–174

Lohoefener R, Lohmeier L (1984) The status of *Gopherus polyphemus* (Testudines, Testudinidae) west of the Tombigbee and Mobile rivers. In: U.S. Fish and Wildlife Service, 126. Report presented in conjunction with a petition to list the Gopher Tortoise west of the Tombigbee and Mobile Rivers as Endangered Species without Critical Habitat, July 15, 1984

McComb WC (1999) Forest fragmentation: wildlife and management implications synthesis of the conference. In: Rochelle JA, Lehman LA, Wisniewski J (eds) Forest wildlife and fragmentation management implications. Brill, Leiden, pp 296–301

McRae AW, Landers JL, Garner JA (1981) Movement patterns and home range of the gopher tortoise. Am Midl Nat 106(1):165–179

Mitchell M (2005) Home range, reproduction, and habitat characteristics of the female gopher tortoise (*Gopherus polyphemus*) in southeast Georgia. M.S. thesis, Georgia Southern University, Statesboro. http://eaglescholar.georgiasouthern.edu:8080/jspui/bitstream/10518/1588/3/Mitchell_Maggie_J_200508_MS.pdf

Mladenhoff DJ, Sickley TA, Haight RG, Wydeven AP (1995) A regional landscape analysis and prediction of favorable gray wolf habitat in the northern Great Lakes region. Conserv Biol 9:279–294

Mushinsky HR, McCoy ED, Wilson DS (1997) Patterns of gopher tortoise demography in Florida. In: Proceedings: Conservation, restoration and management of tortoises and turtles. New York Turtle and Tortiose Society, Bronx, pp 252–258

Ortega-Huerta MA, Peterson AT (2004) Modeling spatial patterns of biodiversity for conservation prioritization in north-eastern Mexico. Divers Distrib 10:39–54

Save Our Big Scrub Inc, Wild South (2006) Petition to list the eastern population of the gopher tortoise as a threatened species. Received January 20, 2006. In Before the Secretary of the U.S Interior and the Director of the USFWS

Simberloff D, Cox J (1987) Consequences and costs of conservation corridors. Conserv Biol 1(1):63–71

Sirakoulis GC, Karafyllidis I, Thanailakis A (2000) A cellular automaton to model for the effects of population movement and vaccination on epidemic propagation. Ecol Model 133:209–223

Smith LL (1995) Nesting ecology, female home range and activity, and population size-class structure of the gopher tortoise, *Gopherus polyphemus*, on the Katherine Ordway Preserve, Putnam County Florida. Bull Fla Mus Nat Hist 38:97–126

Smith RB, Breininger DR, Larson VL (1997) Home range characteristics of radiotagged gopher tortoises on Kennedy Space Center, Florida. Chelonian Conserv Biol 2(3):358–362

Spatialworks (2004) Potential models, tools and approaches for developing habitat objectives to conserve biodiversity in the agricultural regions of Canada. Draft Version 2.0. The Biodiversity Standards Project of the Biodiversity Thematic Group, National Agri-environmental Standards Initiative, Eastern Ontario Model Forest, Natural Resources Canada. Kemptville

Taylor RW (1982) Human predation on the gopher tortoise (Gopherus polyphemus) in north-central Florida. Bulletin of the Florida State Museum. Biol Sci 28:79–102

Tigas LA, Van Vuren DH, Sauvajot RM (2002) Behavioral responses of bobcats and coyotes to habitat fragmentation and corridors in an urban environment. Biol Conserv 108:299–306

U.S. Geological Survey (2003a) A GAP analysis of Georgia: a geographic approach to planning for biological diversity (Final Report 8/10/2003). The Georgia Gap Analysis Project. U.S. Geological Survey, Athens

U.S. Geological Survey (2003b) NLCD land cover class definitions. U.S. Geological Survey. http://www.epa.gov/mrlc/classification.html

van Daalen CE, Dresen L, Janssen MA (2002) The roles of computer models in the environmental policy life cycle. Environ Sci Policy 5(3):221–231

Wilensky U (1999) NetLogo. Computer software. Northwestern University, Center for Connected Learning and Computer-Based Modeling, Evanston. http://ccl.northwestern.edu/netlogo/

Wilson DS (1991) Estimates of survival for juvenile gopher tortoises, *Gopherus polyphemus*. J Herpetol 25(3):376–379

Wilson DS, Mushinsky HR, McCoy ED (1994) Home range, activity, and use of burrows of juvenile gopher tortoises in central Florida. In: Bury RB, Germano DJ (eds) Biology of North American Tortoises. U.S. Fish and Wildlife Service, Fish and Wildlife Research 13, Washington, pp 147–160

With KA (1999) Is landscape connectivity necessary and sufficient for wildlife management? In: Rochelle JA, Lehman LA, Wisniewski J (eds) Forest wildlife and fragmentation management implications. Brill, Leiden, pp 97–115

Wolfram S (1984) Cellular automata as models of complexity. Nature 311:419–424

Wright JS (1982) Distribution and population biology of the gopher tortoise, *Gopherus polyphemus*, in South Carolina. Thesis. Clemson University, Clemson

Wu J, Huang J, Han X, Xie Z, Gao X (2003) Three-gorges dam-experiment in habitat fragmentation. Science 300:1239–1240

Chapter 11
An Individual-Based Model for Metapopulations on Patchy Landscapes-Genetics and Demography (IMPL-GD)

Jennifer L. Burton, Richard F. Lance, James D. Westervelt, and Paul L. Leberg

11.1 Background

Many threatened, endangered, and at-risk species are habitat specialists (Owens and Bennett 2000; Korkeamäki and Suhonen 2002; Munday 2004) that occupy noncontiguous parcels of favorable habitat within matrices of otherwise-unsuitable landscape (Brown 1984; With and Crist 1995). Land managers who are stewards of these species may be faced with difficult decisions about the allocation of conservation resources among parcels, or on which parcels to focus conservation resources in the face of potential habitat losses due to commercial development or changing land-use priorities. Such choices may be critical if the contribution of each parcel to metapopulation viability depends on unfixed attributes of the parcels and the populations that occupy them. If not all parcels contribute equally to the stability of the metapopulation, then the parcels should be prioritized on the basis of conservation goals.

There is a considerable body of literature dealing with identifying important habitat parcels based on multispecies considerations, such as protecting biodiversity

J.L. Burton (✉)
Department of Natural Resources and Environmental Sciences, University of Illinois,
W-503 Turner Hall, 1102 South Goodwin Avenue, Urbana, IL 61801, USA
e-mail: jlburton@illinois.edu

R.F. Lance
Environmental Laboratory, U.S. Army Engineer Research
and Development Center, 3909 Halls Ferry, Vicksburg, MS 39180, USA

J.D. Westervelt
Construction Engineering Research Laboratory, US Army Engineer Research
and Development Center, Champaign, IL, USA
e-mail: james.d.westervelt@usace.army.mil

P.L. Leberg
Department of Biology, University of Louisiana,
300 E Street Mary Boulevard, Lafayette, LA 70504, USA

J.D. Westervelt and G.L. Cohen (eds.), *Ecologist-Developed Spatially Explicit Dynamic Landscape Models*, Modeling Dynamic Systems,
DOI 10.1007/978-1-4614-1257-1_11, © Springer Science+Business Media, LLC 2012

(Reyers et al. 2002; Wiersma and Urban 2005; Rothley 2006). In many cases, however, land managers must focus on a particular species and other land use objectives in addition to conservation. In these circumstances, land managers have a critical need to understand the role of habitat parcels in promoting robust populations of a target species.

11.2 Objective

Our interest is in identifying parcel attributes that make significant contributions to overall metapopulation robustness, and quantitatively incorporating them into a utility index framework that may be used to prioritize parcels for habitat management purposes.

We hypothesized that the risk of metapopulation extinction increases when a habitat parcel is eliminated from a patchy landscape, and that the increase in extinction risk is related to the genetic, demographic, and network characteristics of the removed parcel. We developed the Individual-Based Model for Metapopulations on Patchy Landscapes-Genetics and Demography (IMPL-GD) model to help quantify essential correlations by making it possible to rapidly run large numbers of simulations to test the impacts when habitat parcels are removed.

11.3 Model Description

11.3.1 Purpose

We developed the IMPL-GD model using NetLogo 4.0.4 (Wilensky 1999).[1] The purpose of the model is to support the running of thousands of experiments in which habitat parcels are removed from randomly generated landscapes. After each simulation, ecological variables of the removed parcel—such as physical characteristics of the landscape, population genetic and demographic traits, and network relationships to other habitat parcels—are captured as independent variables along with the impact of parcel removal on the population. Statistical analysis of these experimental results (not covered in this chapter) allows the construction of multiple-regression models that amalgamate habitat and population data into a single conservation utility index (CUI). This index can be used to rank parcel utility relative to a given set of conservation objectives, and to determine which additional data may be most crucial to collect from the field when conservation resources are limited. Once a species-specific algorithm is produced by IMPL-GD, it may be applied to any location to determine which additional critical field data may be needed for management decisions when limited resources preclude a comprehensive study.

---

[1] An operational copy of this model is available through http://extras.springer.com.

11.3.2   State Variables and Scales

The model places generic organisms (whimsically referred to as "whatsits") on a landscape of discrete habitable areas that are separated by traversable but non-habitable terrain. For this study, whatsits were designed to reflect small, solitary animals that defend small, circular territories against others of the same sex. Landscape patches (i.e., cells) are designated either habitable or non-habitable, with each habitable patch identified as part of a discrete habitat parcel. Agents are described by a unique identification number, age, sex, lineage, a binary indicator of heterozygosity, and genetic markers.

The IMPL-GD landscape consists of a user-designated number of discrete habitable parcels, and it wraps horizontally and vertically to avoid edge effects. Because a generic cell is the unit of spatial scaling within the model, the physical landscape and whatsit behavior may be scaled to fit any real-world or hypothetical scenario (refer to the Appendix for a complete description of user-specified settings and ranges).

A single cycle comprising birth, migration, and death represents one time step. While this increment is easily conceptualized as annual, it could be applied to breeding cycles of any length.

11.3.3   Process Overview and Scheduling

The elimination of a habitat parcel occurs at the beginning of the annual time step for which it is scheduled. Whatsit ages are then updated. The generation of new offspring is next, followed by the application of demographic mortality. The time step concludes with whatsit dispersal.

11.3.4   Design Concepts

11.3.4.1   Emergence

For whatsit populations, emergent properties include genetic profiles, extent of inbreeding, and dispersal percentages and outcomes. Nondispersal-related mortality is based on a combination of terrain quality and random factors. Population sizes and rates of change, and migration rates and routes, derive continuously from a combination of landscape features, agent distribution and behavior, and random factors. The genetic properties of populations reemerge with every time step. Along with the magnitude, number, and importance of dispersal routes, population genetic properties influence the importance of parcels within the dispersal network and determine the contribution of each parcel to metapopulation composition, connectivity, and extinction risk. Examples of emergent network and genetic properties that are measured on a global scale in our model include global and average diversity, allelic richness, and total node-degrees.

11.3.4.2 Adaptation

A whatsit will disperse if its patch becomes uninhabitable, or in the step before it becomes sexually mature, or if it is displaced by an older whatsit.

11.3.4.3 Fitness

Individual mortality risk depends on the mean quality (i.e., habitability) of patches within an individual's territory.

The survivability of an individual's offspring (expressed as an adjustment to litter size) depends on the degree of relatedness between that whatsit and its potential mate(s), expressed as Wright's inbreeding coefficient (Wright 1922).

11.3.4.4 Sensing

In each time step, each whatsit reassesses whether the patch it occupies is still habitable. When a whatsit lands on a patch during dispersal it can tell whether that patch is already included in another whatsit's territory, and it senses the sex of the associated whatsit. A female whatsit can breed with any reproductive male, but will not produce any offspring if there is no eligible mate within the user-specified breeding radius.

11.3.4.5 Interaction

Whatsits establish circular territories of a user-settable radius that may encompass both habitable and non-habitable patches. Same-sex territory overlap is optionally prohibited. If a dispersing whatsit settles in a territory already occupied by another whatsit of the same sex, the younger of the two occupants is displaced and will commence dispersal.

11.3.4.6 Stochasticity

Stochasticity drives initial landscape parcel geometry and whatsit distribution. Individual whatsit dispersal and demographic mortality, litter size, and paternity are also constrained random values, based on user-settable variables.

11.3.4.7 Observation

We achieve model validation and data analysis through a combination of real-time visual output and electronic reports. Generation of initial landscapes and subsequent

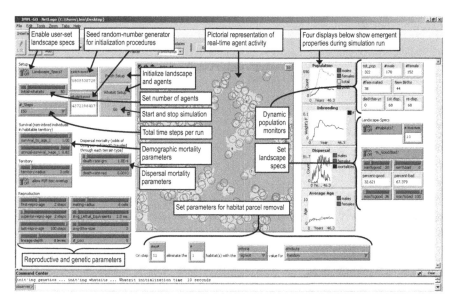

Fig. 11.1 Annotated Individual-Based Model for Metapopulations on Patchy Landscapes-Genetics and Demography (IMPL-GD) user interface

agent activity can be followed directly on a graphical interface, while on-screen monitors provide statistical information in real time (see Fig. 11.1). Reporting functions capture landscape and demographic measurements for each parcel at specified timepoints, while migration and gene flow events are continuously recorded in .csv format files for all network connections between parcels. Additional outputs may be specified by the user.

11.3.5   Initialization

Each simulation began with the random generation of a landscape. Habitat qualities were established by an algorithm that designates 105 random patches to seed the landscape. Thirty-three percent of these were randomly designated as being habitable, and the rest were designated non-habitable. These designations propagated to remaining patches via an algorithm that randomly selects a patch adjacent to one that already had a habitability rating, then gives the new patch the same rating.

Eighty founder whatsits were created to begin each simulation, and unique identification numbers were assigned in series. Male or female sex was randomly chosen for each with equal odds. Founders did not have lineages. All founders were heterozygotes, and each was assigned unique genetic markers. In order to help the model rapidly achieve a steady-state age distribution, all founders were 1 year old at initialization.

11.3.6   Input

Random seeds generated by the model for landscape initialization and initial whatsit distribution were recorded with datasets.

11.3.7   Submodels

11.3.7.1   Habitat Parcel Elimination

If the current time step has been designated for habitat elimination (e.g., step 51 in the experiments performed here), a random habitat parcel is identified, and the quality of patches within that parcel is changed from habitable to non-habitable.

11.3.7.2   Generation of New Offspring

The following steps are executed for each female at least 2 years old. First, breeding males (at least 2 years old) within the mating radius are identified. If there are none, the female will not produce offspring. If breeding males are found within the mating-radius, the number of offspring for this female is determined by the following equation:

$$S * e^{(F*L)}$$

where S is the number of individuals expected to survive to age 1; F is a calculated Wright's inbreeding coefficient; and L is the average number of lethal equivalents.

For each offspring, the father is selected from eligible breeding males within the user-specified mating-radius. Genetic markers are randomly chosen from each parent.

11.3.7.3   Demographic Mortality

For each whatsit at least 1 year old, the probability of dying is a function of the habitable area within that whatsit's territory and a user-settable whatsit survival rate.

11.3.7.4   Dispersal

All year-old whatsits and all individuals displaced by habitat degradation or by other whatsits will disperse. Whatsits begin dispersal in order of age, with the oldest disperser first taking one "stride" (distance = one patch width) in a random direction. That whatsit remains a disperser if the patch it lands on is uninhabitable, or if that patch is within the territory-radius of an older whatsit. For all whatsits who remain

Table 11.1 Overview of parameter values for conservation utility index (CUI) experiments

Parameter	Value
initial-whatsits	80
survival_to_age_1	1
annual-survival_>age_1	0.82
death-rate-grn	0.0001
death-rate-red	0.001
territory-radius	3
allow-M/F-terr-overlap	True
first-repro-age	2
superior-repro-age	2
last-repro-age	100
lineage-depth	8
mating-radius	6
Avg_Lethal_Equivalents	2
avg-litter-size	2
#_Loci	5

dispersers, dispersal mortality is determined at the end of each stride, with odds of death weighted by patch quality, as shown in Table 11.1. If the disperser lands on a patch that is habitable and does not overlap an older whatsit's territory, then it will stay and is no longer a disperser. If this whatsit's new territory overlaps that of any younger whatsit, the younger whatsit will become a disperser.

The whatsit then examines all patches within the territory-radius and moves to the patch that offers the optimal territory quality—in other words, the one with the most habitable patches within its territory-radius. During this search for the most habitable patch within its territory-radius, the whatsit cannot move to a habitable patch that would trigger eviction of a younger whatsit. The dispersal procedure is shown in Fig. 11.2.

11.4 Simulation Experiments

We sought to discover whether the risk of metapopulation extinction would increase with the elimination of a habitat parcel from a given landscape, and whether the magnitude of increase in extinction risk might be related to the size (i.e., area) of the parcel removed.

Following initialization, simulations were run for 50 time steps (50 years) with the intent to allow for achievement of steady state. A habitat parcel was removed at step 51, and the simulation was allowed to run an additional 100 time steps. The model was calibrated to approximate 50% extinction after 150 steps in positive controls (in which a random habitat parcel was removed at step 51).

Simulation results suggest that the size of the parcel that was eliminated had a significant effect on metapopulation extinction. Here, we are concerned with the

204 J.L. Burton et al.

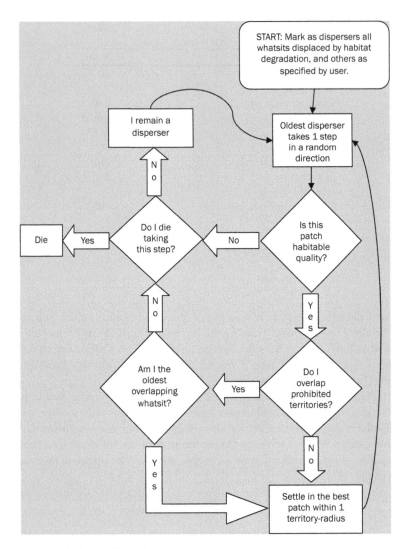

Fig. 11.2 Whatsit dispersal behavior

percentage of populations surviving until the habitat parcel is removed, then becoming extinct within the following 100 steps. If a random parcel was removed, the metapopulation became extinct about 45% of the time (Fig. 11.3). If the smallest parcel was removed, there was a small decrease in the probability of extinction. However, when the largest habitat parcel was removed, the probability of metapopulation extinction increased to 84%. Similar results were obtained from analysis of the number of generations prior to extinction. If the smallest parcel was eliminated, the mean time to extinction was 70 time steps. For random eliminations, the mean time to extinction was slightly lower. However, when the largest parcel was eliminated, the mean time to extinction was 50 time steps.

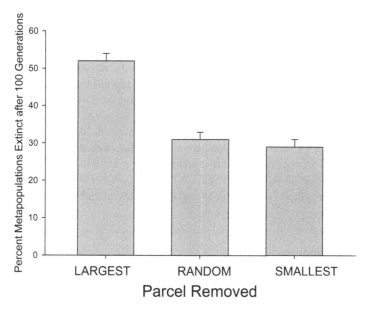

Fig. 11.3 Percentage of simulated metapopulations that became extinct within 50 generations following removal of one parcel of suitable habitat. Parcel selection for removal was based on area (largest or smallest), or the parcel was selected at random

11.5   Discussion

The importance of large habitat parcels to population viability, as shown in these simulations, is obvious and not surprising. Larger habitable parcels are likely to host more individuals, including colonizers and dispersing breeders. Changing a large parcel of suitable habitat into unsuitable habitat can substantially increase the time some dispersers must spend in unfavorable habitat, consequently increasing the proportion of dispersers that die during dispersal. Thus, the loss of a large habitat parcel is likely to have a greater impact on metapopulation viability than the loss of a small parcel. It is of particular interest that removal of a random parcel resulted in an only slightly higher extinction rate than removal of the smallest parcel. This suggests that land managers should focus their conservation resources on the largest parcels, and that saving parcels of intermediate size is unlikely to be significantly more beneficial than preserving the smallest parcels.

Undoubtedly, the spatial arrangement of parcels across a landscape would influence the results of simulations. Likewise, other demographic and ecological factors associated with parcels likely play a role in population viability. When combined, these relative predictors of metapopulation viability could provide a firm basis for supporting management decisions that may involve habitat degradation or fragmentation. This information would be highly valuable to decision-makers who must consider multiple—sometimes even competing—land management objectives. With an improved understanding of the relative importance of habitat characteristics, managers could better utilize existing data and better allocate limited resources to collect the evidence most crucial to determining the relative conservation utility of various habitat areas on a patchy landscape.

Running IMPL-GD thousands of times produces a data set that aggregates the characteristics of each parcel randomly removed during each simulation as well as the population impact of removing each parcel. Using the former as a set of independent variables and the latter as the dependent variable, this data set can be processed using multilinear regression analysis to support the calculation of any habitat parcel's CUI value. This value is a number that indicates the importance of a habitat parcel to species population viability. The experiments described here demonstrate the use of IMPL-GD to identify associations between parcel characteristics and quantify their relationship to species population viability.

11.6 Conclusions

IMPL-GD was designed to assist in understanding how landscape and population characteristics, including network relationships between habitable areas, contribute to metapopulation viability. Through the simulations, we quantified correlations between several relevant demographic, genetic, and network characteristics of habitat parcels. These correlations can provide sound footing on which habitat managers can base decisions that attempt to achieve multiple objectives, including conservation-oriented management of patchy landscapes.

Subsequent to the simulations described here, the authors identified statistically significant correlation between parcel characteristics and the impact of removing a parcel. That analysis falls outside the scope of this chapter, and the results of it are being prepared for documentation in a peer-reviewed journal.

11.7 Appendix

User-specified setting	Description	Range or options
patch-seed	Seeds the random number generator for initial placement of seed patches, to ensure that runs are replicable	Any number
Landscape_Specs?	If true, user-defined parameters constrain acceptance of randomly generated landscapes	True or false
#Habitats?	If true, acceptance of randomly generated landscapes is based on the number of discrete habitat parcels	True or false

(continued)

User-specified setting	Description	Range or options
seed-patches	The number of patches that will be designated "habitable" or "not habitable" to begin landscape initialization	0–350 Patches
seed-good	The percentage of seed patches that will be designated "habitable"	Any whole number percentage
#-Habitats	The number of discrete habitat parcels that will comprise an acceptable landscape	Any whole number
Parcel_Size?	If true, acceptance of randomly generated landscapes is constrained by the area of habitat parcels	True or false
min-terr/hab	The minimum number of whatsit territories available in each acceptable habitat parcel	Any whole number
max-terr/hab	The maximum number of whatsit territories available in each acceptable habitat parcel	Any whole number
%_Good/Bad?	If true, acceptance of randomly generated landscapes is constrained by the proportions of habitable and non-habitable patches	True or false
min%good	The minimum percent habitable patches in an acceptable landscape created via the "seed-all" algorithm	Any whole number percentage
max%good	The maximum percent habitable patches in an acceptable landscape created via the "seed-all" algorithm	Any whole number percentage not > (min%good)
min%bad	The minimum percent non-habitable patches in an acceptable landscape created via the "seed-all" algorithm	Any whole number percentage
max%bad	The maximum percent non-habitable patches in an acceptable landscape created via the "seed-all" algorithm	Any whole number percentage not > (min%bad)
whatsit-seed	Seeds the random number generator for initial placement of whatsits, to ensure that runs are replicable	Any number
initial-whatsits	The number of founder whatsits created to begin the run	Any whole number
#_Steps	The total number of steps included in the replicate	To extinction 1 10 25 50 100 150
step#	The step on which habitat elimination will occur	Any whole number

(continued)

User-specified setting	Description	Range or options
#	The number of habitat parcels that are eliminated at the scheduled step	Any whole number
criteria	Determines the order in which habitat parcels are selected for elimination	Highest or lowest
attribute	The characteristic by which habitat parcels are chosen for elimination	Random #-of-whatsits Births-previous-year Deaths-previous-year Net-population-change-previous-year #-outmigrants-previous-year Area Edge-to-area-ratio Distance-to-closest-habitat Female:male_ratio #-founders-represented #-founders-represented-only-here #-alleles #-private-alleles Heterozygosity Abs-diversity Relative-diversity Differentiation-relative Contribution-to-total-diversity Allelic-richness Contribution-to-total-allelic-richness Node-degree Connectivity-contribution Bonacich's-centrality Node-strength Sum-of-linked-nodes'-strengths Contribution-to-connectivity-strength #-of-founders-represented
survival_to_age_1	The mortality rate for individuals in the first year of life	0–1 (increment 0.001)
annual-survival_>age_1	The mortality rate for individuals at least 1 year of age	0–1 (increment 0.001)
death-rate-grn	The odds that a dispersing whatsit will die while traveling one cell-length through habitable terrain	0 to (death-rate-red) (increment 0.0001)
death-rate-red	The odds that a dispersing whatsit will die while traveling one cell-length through non-habitable terrain	(death-rate-grn) to 1 (increment 0.0001)

(continued)

User-specified setting	Description	Range or options
territory-radius	The distance (in patches) from a whatsit's location to the boundary of its territory in all directions	1–20 Cells (whole numbers)
allow-M/F-terr-overlap	This setting determines whether or not a patch may be included simultaneously in both a male and a female territory	True or false
first-repro-age	The minimum age at which whatsits are capable of reproducing	0–5 Steps
superior-repro-age	The age at which individuals become more likely to reproduce than younger whatsits	0–25 Steps
last-repro-age	The maximum age at which whatsits are capable of reproducing	(superior-repro-age) to 100 steps
lineage-depth	The number of generations contained in each whatsit's lineage	0–20 Levels
mating-radius	The maximum distance (in patches) at which whatsits are able to mate	1–10 Cells (whole numbers)
Avg_lethal_equivalents	The average number of lethal equivalents for the population	0–20 (Increment 0.1)
avg-litter-size	The minimum average litter size	0–20
#_Loci	The number of loci for which genetic information is recorded	0–449

References

Brown JH (1984) On the relationship between abundance and distribution of species. Am Nat 124(2):255–297

Korkeamäki E, Suhonen J (2002) Distribution and habitat specialization of species affect local extinction in dragonfly Odonata populations. Ecography 25(4):459–465

Munday PL (2004) Habitat loss, resource specialization, and extinction on coral reefs. Glob Chang Biol 10(10):1642–1647

Owens IPF, Bennett PM (2000) Ecological basis of extinction risk in birds: habitat loss versus human persecution and introduced predators. Proc Natl Acad Sci U S A 97(22):12144

Reyers B, Fairbanks DHK, Wessels KJ, Van Jaarsveld AS (2002) A multicriteria approach to reserve selection: addressing long-term biodiversity maintenance. Biodivers Conserv 11(5):769–793

Rothley KD (2006) Finding the tradeoffs between the reserve design and representation. Environ Manage 38(3):327–337

Wiersma YF, Urban DL (2005) Beta diversity and nature reserve system design in the Yukon, Canada. Conserv Biol 19(4):1262–1272

Wilensky U (1999) NetLogo. Computer software. Center for Connected Learning and Computer-Based Modeling, Northwestern University, Evanston. http://ccl.northwestern.edu/netlogo/. Accessed 01/2011

With KA, Crist TO (1995) Critical thresholds in species' responses to landscape structure. Ecology 76(8):2446–2459

Wright S (1922) Coefficients of inbreeding and relationship. Am Nat 56:330–338

Chapter 12
An Implementation of the Pathway Analysis Through Habitat (PATH) Algorithm Using NetLogo

William W. Hargrove and James D. Westervelt

12.1 Background

When developing plans to protect populations of species at risk, we tend to focus on protecting and preserving habitat where the species naturally occurs. Ecologists employ historic sightings of individuals, habitat suitability index (HSI) models, and the expert advice of local naturalists familiar with the environments desired and required. In many cases, the optimal habitats are fragmented to the extent that a species at risk could be extirpated from any single area by storms, drought, disease, or other local insult. Alternately, over time, a small population could become so inbred that it becomes susceptible to disturbances that increase the probability of local extirpation. However, if nearby populations can reach the isolated population across the fragmented space, then the isolated population becomes part of a larger metapopulation. The strength of habitat connections, therefore, can be critical to ensure the viability of at-risk populations over time. Habitat connectivity plays this important role by increasing the effective population size, maintaining gene flow, and facilitating regular migration, dispersal, and recolonization. Each of these processes helps ensure the long-term persistence of a population. A connected landscape is preferable to a fragmented one (Beier and Noss 1998; Bennett 1999). Natural landscapes are generally more connected than landscapes altered or manipulated by humans, so establishing or maintaining corridors is a viable strategy to sustain the natural connectivity (Noss 1987).

W.W. Hargrove (✉)
USDA Forest Service Southern Research Station, Eastern Environmental Threat
Assessment Center, 200 W.T. Weaver Blvd., Asheville, NC 28804-3454, USA
e-mail: whargrove@fs.fed.us; hnw@geobabble.org

J.D. Westervelt
Construction Engineering Research Laboratory, US Army Engineer Research
and Development Center, Champaign, IL, USA
e-mail: james.d.westervelt@usace.army.mil

J.D. Westervelt and G.L. Cohen (eds.), *Ecologist-Developed Spatially Explicit*
Dynamic Landscape Models, Modeling Dynamic Systems,
DOI 10.1007/978-1-4614-1257-1_12, © Springer Science+Business Media, LLC 2012

Metapopulation is the term for a collection of discrete local breeding populations that occupy distinct habitat patches but are connected by migration (Hanski and Gilpin 1997). Population viability analysis (PVA), a method for forecasting the probability that a metapopulation will persist over time, has been automated by numeric modeling applications such as RAMAS (Applied Biomathematics 2003). A key input to a PVA is the probability of migration among all individual populations of the metapopulation. Understanding the viability of an at-risk population or metapopulation depends to a significant extent on identifying the migration routes that connect the discrete constituent populations. Also, without this understanding, land managers may not have enough information to protect important migration paths. That deficiency, in turn, may lead to the loss of more-isolated populations, an increased threat to the survival of the greater metapopulation.

In order to help improve our understanding of species migration routes among separated populations, Hargrove, Hoffman, and Efroymson (2005) developed the Pathway Analysis Through Habitat (PATH) computer simulation model. Originally developed for implementation on a supercomputer, the purpose of PATH is to help a decision-maker to reliably predict where potential dispersal corridors are likely to exist in real-world landscape maps. This information makes it possible to project which habitats will support population growth (sources) and which will tend to lose population (sinks). The PATH algorithm works by launching *walkers* (i.e., virtual animals) from each habitat patch to simulate the journey of individuals through land cover types in the intervening matrix until arriving at a different habitat patch or dying. Each walker is given a set of user-specified habitat preferences that direct its walking behavior to resemble the animal of interest. As originally implemented, PATH was designed for a massively parallel computing environment in order to analyze the activity of very large numbers of random walkers in large landscapes with many habitat patches. PATH produces three outputs: (1) a map of the most heavily traveled potential migration pathways between patches, (2) a square transfer matrix that quantifies the flow of animals successfully dispersing from each patch to every other patch, and (3) a set of importance values that quantifies, for every habitat patch in the map, the contribution of that patch to successful animal movement across the landscape. The transfer matrix is square and not triangular since the rate of animal movement is likely to be asymmetrical between any two habitat patches. That is, the rate of successful migration from patch 1 to patch 2 will likely not equal the successful migration from 2 to 1.

One problem with the original implementation of PATH is the inherent barrier to its use: the application is designed to perform massive simulations that require a supercomputer to run them in a reasonable amount of time. This project was conceived as a way to make migration pathway analysis more accessible to users of desktop computers by implementing the core PATH algorithm in the NetLogo simulation modeling environment (Wilensky 1999).

12.2   Objective

The objective of this modeling project was to develop a computationally efficient PATH-based tool for desktop computers that can identify important species migration corridors between habitats based on expert information about inter-habitat patch lethality, the energy cost to cross, and the energy available to animals to move outside of core habitat.

12.3   Model Description

12.3.1   Purpose

This implementation of PATH is intended to illuminate the essential mechanisms that help to identify animal migration corridors.[1] A premise of the design was to avoid including any data or processes not directly illuminating the successful migration of species between two or more separated habitat patches. This model converts expert knowledge about habitat patch locations, traversal cost, and probability of mortality through the interstitial landscape into information about the relative connectivity of all pairs of habitat patches and the impact of interstitial lands on successful migrations. In future versions, certain additional aspects of "realism," such as seasonal effects and time steps, could be added in instances where model results might be improved. As you read further you will note that, unlike the other models in this book, this model does not use time steps.

12.3.2   State Variables and Scales

The scale and extent of the area are determined by the user, with the practical limit depending on the processing power of the user's computer. The demonstration data set used in this model supports a simulation space encompassing more than 1,000 by 1,000 cells (i.e., one million cells). The primary state variable that changes over simulation time is the number of successful migrations supported by each patch in the simulation space.

---

[1] An operational copy of this model is available through http://extras.springer.com.

12.3.3 Process Overview and Scheduling

The model is initialized by reading data from a set of three location-specific maps that characterize (1) the location of habitat patches, (2) the energy cost to cross the space between patches, and (3) the lethality associated with migrating through non-habitat. During initialization, first the habitat map loads, then the model aggregates, outlines, and numbers all contiguous habitat patches (i.e., "habitats"). A user's chosen number of travelers is initialized. The simulation runs in discrete steps, but these steps do not represent time because time has no bearing on an individual's energy level, the energy cost of migrating, or patch lethality. At each step in the simulation, walkers are randomly distributed along the edges of a habitat based on area size. Walkers are faced away from the habitat interior and started on a walk that is partially directed with an adjustable level of randomness. The amount of energy consumed during one step is based on the data provided by the energy cost input map, and travelers randomly die according to the patch-specific probability of mortality read from the patch lethality input map. As a traveler moves through the interstitial space between habitats, it remembers its course. If the traveler succeeds in arriving alive at a habitat patch different from where it started, it communicates to every patch along the successful path that the patch supported a successful crossing. Each patch then updates a habitat-to-habitat crossing array that tracks the number of successful crossings among all patch combinations.

12.3.4 Design Concepts

12.3.4.1 Emergence

PATH reveals the value of every patch in supporting habitat-to-habitat migration through the emergent behavior of the walkers, as an increasing number of them successfully complete migrations and the most favorable paths become more evident.

12.3.4.2 Stochasticity

Walkers begin their attempted migrations by facing away, randomly, from the interior of their beginning habitats, and then proceed to move through the space separating habitat patches. The travel direction of walkers may be set to be fully random, partially random, or fully deterministic depending on the value selected for a user-set variable. The movement of a walker at each simulation step may be characterized as a "wiggle," turning to the left between 0 and X angular degrees (where X is a user-selected value); then turning to the right between 0 and X degrees. If the user sets X to 0, then the travel path is straight; if the angle is set to 360, then every step is fully random. The user will assign a value to X based on what is known of the subject species' tendency to maintain a direction; movement tendencies vary among species.

12.3.4.3   Observation

In this version of the model, walkers observe nothing about their surroundings because that information is largely extraneous to path selection on a collective level. However, this model could be modified for experimentation purposes to support species-specific observation and evaluation of travel-direction options by individual walkers.

12.3.4.4   Time

This model deals with time in a manner very different from any other model in this book, which all include an idea of time passing as the model executes. In this PATH model, each step involves releasing a new batch of walkers, which expend energy and risk death as they move. One may imagine that time passes as they move, but their "time" is not associated with any other walker's "time" in any model step. Regardless of the number of ticks, the total set of walkers active during the simulation can be assumed to begin their walking at the same time. Also, the state of the landscape never changes as it often does in other models, which leaves the landscape essentially timeless.

12.3.5   Initialization

Patches are initialized with data read from the habitat-location, energy-cost, and patch-lethality maps. Habitat quality is represented by a binary variable, either 0 for non-habitat or 1 for habitat. The walkers' energy store is set by the user before migration attempts begin, and the amount of energy lost to walkers as they cross patches is accounted for by data from the model's energy-cost map. Finally, the probability of mortality while crossing a patch is represented by a value ranging from 0 to 100 as determined by the model's lethality map. Also during initialization, as briefly noted above, patches are clumped together to form contiguous habitats, and each habitat is given a unique identification number.

12.3.6   Input

Input is provided by the three raster maps described above. The habitat-location, energy-cost, and patch-lethality maps are prepared within a raster-based geographic information system (GIS) and provided to the model as Esri ASCII grid files. PATH requires maps that are location- and species-specific. The habitat map could be produced using the results of a Habitat Suitability Index study. It also might be based on a regression analysis that correlates known habitat and non-habitat patches with factors in other GIS maps such as slope, land cover, soil type, land use, canopy cover, and elevation. In any case, the end product must be a map of habitat for the target species—places where the species can establish a home range and survive. The lethality map represents the probability (0–100%) that an individual will die while crossing any patch. A patch lethality value is based on exposure to predators, the probability of being caught in an inescapable situation, and the species' ability to deal with the land cover. GIS data involved in developing this information might include land cover, land use, slope, and aspect. The energy-cost map is similar to the lethality map, but instead of encountering immediate death in specific patches, the walker loses energy based on the energy-cost value for the patch being crossed. For example, crossing through dense woods, a swamp, or a pond may require more energy than crossing an open short-grass field. Energy cost, of course, varies with the species. Like the lethality map, the energy-cost map will likely be based on an ecologist's analysis of land cover, slope, and aspect data.

12.3.7   Submodels

Our NetLogo implementation of PATH sets aside all ecosystem and species functions that are not essential to the modeling of animal migration path formation. It is a very simple model that operates on data provided by three raster maps as initialized with species-specific values set by the user for a small number of variables.

The concept of time is not essential to understanding the establishment of migration paths, so each NetLogo tick (i.e., step) occurs independently of time, and represents only one discrete action that involves walkers attempting to migrate from a home habitat to another habitat. At each NetLogo tick, a user-chosen number of travelers make a crossing attempt. That number is not critical, but the number of total attempts is. The number of travelers leaving each habitat is proportional to the size of the habitat, based on the assumption that larger habitats have larger populations and therefore will send forth more emigrants. Walkers depart home habitats from a random edge location.

As the simulation proceeds, successful migration paths are traced onto the output map in black, and each path becomes denser as more individuals use the same path. These paths accrete into increasingly stable shaded areas that reveal the corridors between habitats that are most successfully used in migration. The darkest traces within these grayscale areas indicate the most successfully used corridors.

12.4   Simulation Experiments

We used this PATH application to identify the corridors within Fort Benning, GA, that might be used by gopher tortoises (*Gopherus polyphemus*) migrating among tortoise habitat fragments. This animal, an at-risk species that has been nominated

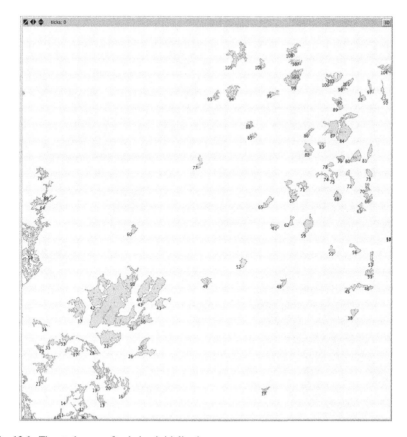

Fig. 12.1 The study area after being initialized

for Federal Threatened status, is carefully managed on Fort Benning. The three GIS maps required for the model were derived from National Land Cover Data (NLCD) maps. Areas with land cover suitable for tortoises were selected, and contiguous areas greater than 18 ha were identified as habitat. Each of the NLCD land cover types was then associated with a transit-energy cost and a probability of mortality (0–100%), and these parameters were then used as the basis for reclassifying the NLCD map into the other two input maps.

Model initialization, which involves reading the three maps as well as identifying and labeling contiguous habitat patches into habitats, takes several minutes and produces the image in Fig. 12.1. The 108 identified habitats are shown in medium gray, with edges outlined in darker gray and uniquely labeled. For this experiment, each walker was given 4,000 units of energy before its crossing attempt and was assigned a maximum turn angle of 20° after each step. Five hundred walkers were arbitrarily initialized at the same time, started at a habitat edge and randomly faced away from the habitat. The total leaving any habitat was based on that habitat's relative size. Each walker moved ahead until it ran out of energy or successfully reached

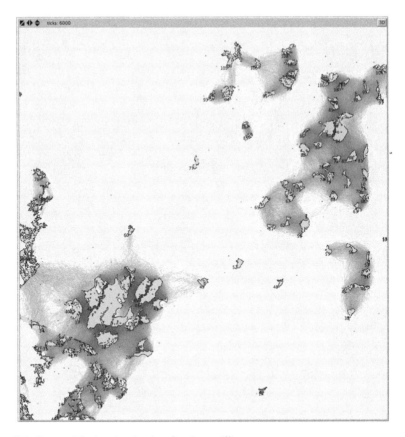

Fig. 12.2 Successful migration density after three million attempts

a new habitat. Once these walkers were finished, a new set of 500 was initialized and the process continued over and over. In 1 min of "wall-clock time," the model generated 1.25 million walker attempts, with about 10% successfully migrating from 1 of the 108 habitat patches to another. After three million habitat-crossing attempts were simulated, the patches were assigned a color along a log-adjusted gray scale based on the relative number of successful crossings in which they were used (Fig. 12.2). The specks in the image identify locations at which travelers ran out of energy in their final attempt. The maximum number of successful crossings supported by any one patch was 2,932 and is represented in black. Numerous patches participated in crossing. When the simulation is viewed on a computer screen, these patches are visible as white sinuous lines against the darker background.

Just 357,328 of the 3 million crossing attempts were successful. This value was derived from the inter-habitat migration success table, which can be displayed by selecting NetLogo's Interface tab. For this simulation, four strong metapopulations emerged, with three of them being tenuously linked (inspect Fig. 12.2). With lower initial energy levels, the number of successful migrations decreased, the number of

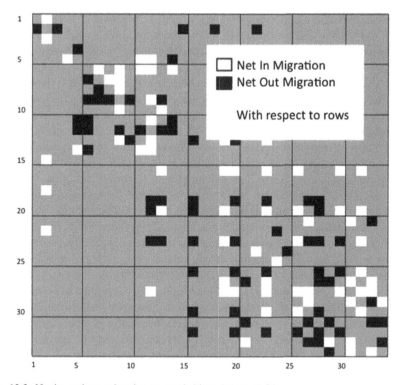

Fig. 12.3 Net in- and out-migration among habitats 1 through 34

separated metapopulations increased, and the number of habitats unconnected to other habitats increased. In terms of species conservation and survival, higher numbers of metapopulations and disconnected habitats are less desirable; and higher numbers of individuals and larger metapopulations are more desirable.

One output of this application is a table that shows the number of successful migrations between all pairs of habitats. By subtracting out-migrations from in-migrations, the user can discover which habitats in each pair are net population sources or sinks. Figure 12.3 shows the net migration from every habitat to every other habitat for habitats 1–34 (out of 108 total). Note that habitat 2 is a net source to habitats 1, 3, 15, 18, and 21 while habitat 16 is a net sink for 7 other habitats.

12.5   Discussion

This PATH application reveals that high-quality interstitial habitat is more likely to be associated with corridors than less-preferred habitat, and that short connectors develop more often than longer ones. The model assumes that movement through non-habitat areas is semi-random, with the walker's current heading having some persistence. Actual animals may functionally select resources at a finer scale than is used in this model, and thus may not follow the same route as one optimized by PATH at a coarser resolution. However, animals also may respond to coarser-grained landscape cues than are represented by this model's habitat map, especially when migrating or dispersing across long distances. Animal movement choices and behaviors may vary with age, gender, pregnancy and nutrition status, and the nature and composition of traveling groups. The importance of those factors may differ between species.

The implementation of PATH described here attempts to illuminate patterns of successful inter-habitat migration with a computationally low-overhead modeling approach. It encapsulates essential migration activities and costs into bare fundamentals—a binary habitat indicator, a movement parameter, a randomness parameter, an energy-accounting function, and a mortality probability. The model also includes output basic functions to track and display successful crossings. The map interface shows only the metapopulations and the strength of population connections within them. This information can be concisely output to reveal whether given populations are likely sources and sinks (Fig. 12.3).

Although this PATH-based model is very simple, it is important to note that it is technically sufficient to capture the essential elements of real-world migration path-formation for a variety of species. While simpler analytical approaches are possible, such as a least-cost method for predicting path formation, they do not account for multiple terrain types that differ in terms of crossing-energy costs or lethality probability, and are too simplistic to capture all essential considerations. More importantly, because least-cost path modeling methods show the solution to be a single pathway, they illustrate only the current optimal route but do not show secondary routes that could be improved over the current best route through small modifications to the habitat map. The PATH tool shows all feasible connectivity routes, not just the single current best one. Paths that have good potential, but are currently used as secondary routes by the metapopulation, are exactly the ones that resource managers need to know about in order to consider where strategic management alterations can dramatically improve habitat connectivity. Such routes are evident in a PATH output corridor map as strong potential corridors, which may be impeded only in one or two locations by passing through low-quality patches.

The walkers in a simulation, considered individually, have almost no sophistication at finding migration paths as compared with individual real-world animals. However, considered collectively, the behavior of all successful walkers represents a spatial optimization process that can be used to reproduce the optimum pathways we would expect well-adapted individuals of the subject species to use most often. For this reason, it is not a problem that the walkers, endowed only with the ability to "see" land types immediately adjacent to their current location, represent animals that may have much greater sensory scope and range. The simplified walker mode of functioning in the model does not constrain the optimization of the potential corridors found by PATH. The optimization in this model results from the collective action of the large number of successful dispersers and the weighting of the most efficient potential dispersal paths the most heavily. Even if individual walkers in the model had a greater look-ahead capability, the same optimal potential corridors would be predicted by the PATH tool. Consider, for example, a situation in which the local environment is inhospitable, but just beyond this there is a high-quality habitat pathway to another patch. Although short-sighted, a few walkers will make rare, immediately suboptimal choices and cut through the inhospitable bottleneck to discover the optimized pathway beyond. Becoming successful dispersers, their tracks will become part of the final map, showing the connection just as surely as if they had been able to look beyond the local problem with terrain. A converse situation is also true: pockets of higher-quality habitat that are surrounded by barriers of poor-quality habitat will attract many individual walkers, but it is not likely that those walkers will successfully reach a second patch. Since only the travel paths of successful dispersers are used, no potential corridors will pass through this attractive dead-end area, just as if walkers had been able to see the barriers to migration that lie beyond.

For purposes of real-world application to habitat-management decisions, corridors identified using randomized walkers in this PATH tool must be verified against actual movement corridors that have been observed in the field. Radio tracking and telemetry studies would be appropriate for empirically testing walker-generated corridors against migration paths established by the subject species in the natural environment.

12.6   Conclusions

The original PATH model, designed to handle very large numbers of walkers on very large landscapes, was implemented on a supercomputer. Few working resource managers actually have access to a supercomputer. This implementation of the PATH algorithm as a simplified NetLogo model makes PATH available to a much wider user community because it works well on a standard desktop computer. This PATH-driven model identifies and highlights areas in a landscape that contribute to the natural connections among populations, identifies the metapopulation structure, and indicates the relative strength of connections holding a metapopulation together. This information is essential to making effective habitat-management decisions that support robust populations of species at risk.

References

Applied Biomathematics (2003) Ramas software. Copyright 1982–2007. Setauket. http://www.ramas.com/. Accessed 2/26/2010
Beier P, Noss RF (1998) Do habitat corridors provide connectivity? Conserv Biol 12(6):1241–1252
Bennett AF (1999) Linkages in the Landscape: The Role of Corridors and Connectivity in Wildlife Conservation. International Union for Conservation of Nature and Natural Resources (IUCN) Forest Conservation Programme. Conserv Forest Ecosystems, ISBN 2831702216, 9782831702216, 24(1):244

Hanski IA, Gilpin ME (eds) (1997) Metapopulation biology. Academic, San Diego

Hargrove WW, Hoffman FM, and Efroymson RA (2005) A practical map-analysis tool for detecting potential dispersal corridors. Landsc Ecol 20(4):361–373

Noss RF (1987) Corridors in real landscapes: a reply to Simberloff and Cox. Conserv Biol 1(2): 159–164

Wilensky U (1999) NetLogo. Computer software. Center for Connected Learning and Computer-Based Modeling, Northwestern University, Evanston. http://ccl.northwestern.edu/netlogo/. Accessed 2/26/2010

Chapter 13
A Technique for Rapidly Forecasting Regional Urban Growth

Todd BenDor and James D. Westervelt

13.1 Background

In 2007, for the first time in history, more than half of the world's population resided in urban areas (United Nations Population Fund 2007). In the USA, the transition from dense urban centers and dispersed rural populations to expansive suburban landscapes has been rapid (Calthorpe and Fulton 2001). When compared with US population growth, this development history has led to a disproportionately high rate of land use change and the creation of concentrated stretches of impervious (impermeable to water) surfaces (Ewing et al. 2002). Studies have observed this expansion, noting the drop in urban densities and rapid expansion of land developed for urban uses and infrastructure (Deal and Schunk 2004; Ewing et al. 2002).

Excessive suburbanization (often referred to as *sprawl*) has concerned researchers and planners, sparking a demand for new tools to help understand the effects of large-scale urbanization. Among these tools are computer-based simulation models of urban growth. Several studies have surveyed these models, describing their advantages, disadvantages, and intended uses (Agarwal et al. 2002; EPA 2000). Agarwal et al. (2002) analyze these models and their capabilities in terms of three considerations: spatial detail, ability to mimic human decision-making capability, and dynamic implementation. However, from the perspective of researchers or planners interested in applying these models, important additional practical factors affecting adoption include the level of technical complexity, cost of implementation,

T. BenDor (✉)
Department of City and Regional Planning, University of North Carolina at Chapel Hill, Campus Box 3140 New East Building, Chapel Hill, NC 27599-3140, USA
e-mail: bendor@unc.edu

J.D. Westervelt
Construction Engineering Research Laboratory, US Army Engineer Research and Development Center, Champaign, IL, USA
e-mail: james.d.westervelt@usace.army.mil

J.D. Westervelt and G.L. Cohen (eds.), *Ecologist-Developed Spatially Explicit Dynamic Landscape Models*, Modeling Dynamic Systems,
DOI 10.1007/978-1-4614-1257-1_13, © Springer Science+Business Media, LLC 2012

magnitude of data requirements, and the amount of time required for model implementation. These models are also difficult to implement for teaching purposes, requiring months or even years to parameterize, populate with data, and run.

Tools for modeling urban environments and land use change have been shown to be useful for predicting and understanding urban, social, and ecological problems (Agarwal et al. 2002; Barredo et al. 2003; Grimm and Railsback 2005). Additionally, these tools have been used to present environmental complexities to wide audiences, thereby establishing their value as educational and visualization tools (Costanza and Voinov 2001, 2004; Ford 1999).

Waddell and Ulfarsson (2004) provide an introduction to the design and development of operational urban simulation models, recommending a number of steps to developing useful modeling tools. These steps include assessments of institutional, political, and technical context; and stakeholders, value conflicts, and public policy objectives. These assessments are followed by development of measurable benchmarks for objectives, inventories of policies to be tested, maps of policy inputs to outcomes, and assessment of model requirements. Finally, input data are prepared, followed by model specification, estimation, calibration, validation and, finally, model usage.

Land use modeling frameworks typically address specific questions, such as:

- Where are people likely to be living within urban regions over the next 20–50 years?
- To what extent and where will new impervious surfaces be introduced into the landscape, and how will these affect nutrient or pollutant input into downstream water bodies?
- What percentage of important natural habitat in the region will be located on publicly owned land?
- How much habitat is likely to be damaged, and in what patterns?

There is a considerable body of literature on urban growth modeling at the local and regional scales, as reviewed in various studies (Agarwal et al. 2002; EPA 2000). Well-known city-scale models include METROPILUS (Putman 1983), MEPLAN (Echenique et al. 1990), and UrbanSim (Waddell 2002). Some of these have been developed to work in conjunction with commercial geographic information system (GIS) software, such as What-if (Klosterman 1999). Regional-scale models typically cover multiple counties and attempt to forecast land development at the edges of and beyond cities. Examples include SLEUTH (Clarke and Gaydos 1998; Jantz et al. 2003), the Land use Evolution and impact Assessment Model (LEAM; Deal and Schunk 2004; Wang et al. 2005), and California Urban Futures-2 (CUF-2; Landis and Zhang 1998). In order to adequately represent the complexities of urban systems, while handling the computational requirements for testing alternative policy scenarios, construction, initialization, and calibration of urban and regional models tend to be very resource-intensive and time-consuming.

In this project we explored the development and application of a much smaller urban simulation model, constructed in NetLogo (Wilensky 1999) working in conjunction with purpose-built input maps of two small US cities. Our hypothesis is that even a small, expedient urban model can teach students and nonspecialists about important cause–effect relationships in urban dynamics.

13.2 Objective

The objective of this project was to provide a classroom-level modeling exercise that allows students to experiment with urban growth concepts.

13.3 Model Description

13.3.1 Purpose

This model helps users to explore and visualize urban growth patterns in response to land cover and urban growth attractiveness maps, neighborhood development, projected development, and randomness. Project development proceeded in two phases. First, raster-based GIS overlay processing and analysis code was written to create urban growth attractiveness based on National Land Cover Data (NLCD) maps for an area of interest. This work is described in Sect. 13.3.6. Second, a spatially explicit simulation model was developed with NetLogo 4.1 (Wilensky 1999) to generate new urban areas based on several user inputs, including growth needed, development of and size of neighborhoods, effect of growth on attracting new adjacent growth, and a user-selected level of randomness in the growth.[1]

The model interface is displayed in Fig. 13.1, which divides the simulation into a three-step process. The model comes with map sets for two locations: Champaign-Urbana, IL, and Chapel Hill, NC. In the first step, the user initializes the model with the map of either location by selecting the appropriate button. The figure illustrates what appears when the "initialize-champaign" button is clicked. In the second step, the user sets four variables (discussed in Sect. 13.4). In the third step, the user designates the length of time, in years, that the simulation will represent. The simulation can be stepped 1 year at a time or run for the full number of years by clicking the "step" or "run steps" button.

13.3.2 State Variables and Scales

The single state variable in this model is a binary value for each patch indicating new urban development. Sets of two input maps are provided with the model for Chapel Hill, NC, and Champaign-Urbana, IL. The resolution of these maps is 30 m. The Champaign-Urbana map covers approximately 20×20 km, and the Chapel Hill map covers 21.5×22.7 km. Development of the maps, land cover, and attractiveness to urban growth are described below. The maps are used to establish two variables for each patch: a value representing the land cover type and a value ranging from

---

[1] An operational copy of this model is available through http://extras.springer.com.

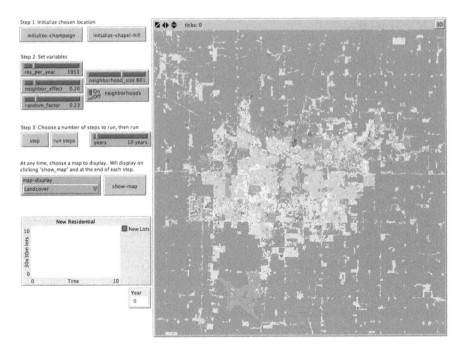

Fig. 13.1 Model interface

0 to 1.0 that represents the attractiveness of each patch to new development. The second value can change during model runs depending on the development of immediate neighbors.

Several input variables are supported by slider bars in the interface:

res_per_year	This number of patches will be converted to urban use each year
neighbor_effect	This 0–1.0 value is used to adjust the development attractiveness of patches immediately neighboring new development. If set to 0, there is no effect on development attractiveness. If set to 1, attractiveness will increase to 1.0
random_factor	The development attractiveness input map values are randomly modified based on this 0–1.0 value. If set to 0, the attractiveness values are not modified; if set to 1.0, they are completely replaced with random values. A blending of these extremes occurs with the selection of other values
neighborhoods	This on/off switch toggles neighborhood production. When on, all new growth will be in the form of neighborhoods that are equal to or less than the neighborhood_size
neighborhood_size	If "neighborhoods" is turned on, then growth will occur in clusters that are no larger than this number

Using these variables, students can explore urban growth concepts.

13.3.3   Process Overview and Scheduling

At each time step, the following events occur:

1. The attractiveness of each patch is updated based on recent new development.
2. Patches are sorted based on attractiveness and a random factor.
3. A user-specified number of patches are selected for development.

13.3.4   Design Concepts

13.3.4.1   Emergence

This urban growth model operates at the patch level, but it generates an emergent pattern of growth in response to the input maps and user selection of variable values.

13.3.4.2   Stochasticity

Random values are employed at two points in the model. First, the attractiveness of each patch to new development is modified with a uniquely generated random value to take into account factors that were not specifically included in the development of the input urban attractiveness map. The weight of that value is user-adjusted via the *random_factor* variable entered through the interface. Depending on the setting, the adjusted attractiveness can be left unchanged or can be completely replaced with the random value. Second, if neighborhood development is turned on, neighborhood growth proceeds randomly around an initially selected seed location.

13.3.5   Initialization

Each patch is initialized with two variables: a land use category based on the 1992 NLCD values; and a base urban growth attractiveness value in the form of a 0–1.0 value index. These values are provided to the model through Arc-ASCII (.asc) grid maps. Growth is allowed to occur in areas not categorized as water, wetland, or already urban (including roadways).

13.3.6   Input

The NLCD map is created by acquiring digital land cover maps and incorporating that information into a map with the 1992 NLCD data categories listed in Table 13.1.

Table 13.1 1992 National Land Cover Data (NLCD) categories used in input maps

Category	Description
11	Open water
12	Perennial ice/snow
21	Low-intensity residential
22	High-intensity residential
23	Commercial/industrial/transportation
31	Bare rock/sand/clay
32	Quarries/strip mines, gravel pits
33	Transitional
41	Deciduous forest
42	Evergreen forest
43	Mixed forest
51	Shrubland
61	Orchards/vineyards
71	Grasslands/herbaceous
81	Pasture/hay
82	Row crop
83	Small grains
84	Fallow
85	Urban recreational grasses
91	Woody wetlands
92	Emergent herbaceous wetlands

The map displayed through NetLogo on the right-hand side of Fig. 13.1 shows the sample Champaign-Urbana area. The same data for other cities can be directly downloaded from the United States Geological Survey (USGS) "seamless" website: http://seamless.usgs.gov.

The urban growth attractiveness map provides an index value in the range of 0–1.0, representing the relative attractiveness of every raster GIS grid cell to new development. We developed the sample maps using a *hedonic logic* approach, which is typically used to estimate land values in the resource economics literature. The logic calculates and then combines the level of attractiveness of each grid cell with respect to (1) surrounding urban area centers, (2) access to roads, highways, and interstates, and (3) proximity to forest and water. The analysis process is accomplished through five primary steps:

1. Acquire data and re-project it into a common coordinate system.
2. Identify a set of locations for each attractor believed to influence development.
3. Calculate travel times to each attractor forming a map.
4. Convert the travel time maps to attractiveness maps.
5. Combine attractiveness maps to generate a single, comprehensive urban development attractiveness map.

Steps 1–4 are accomplished for each landscape attractor. The results for the attractors are combined in step 5 using a simple binary logistic regression model, shown in (13.1):

$$A = \beta_0 + \sum_{i=1}^{k} \beta_i V_i, \quad P = \frac{e^A}{1+e^A} \tag{13.1}$$

where:

A = Overall attractiveness of a parcel to development
β_0 = Y-axis intercept of the regression
β_i = Fit by Maximum Likelihood Estimation (MLE; binary or multinomial logistic regression)
V = Value of attractor (k values)
P = Probability of urban growth occurring on cell

A more detailed description of this process can be found in Westervelt et al. (2011).

13.3.7 Model Logic

The steps at each time step (year) proceed as follows. First, an urban attractiveness value is calculated for each patch that has not already been developed, was not originally urban, and is not classified as water or wetland, based on the following equation:

$$D = F * R + (1.0 - F) * A \tag{13.2}$$

where:

D = Development probability index
F = User chosen "random_factor"
R = A random number unique to each patch (0–1.0)
A = Base development probability index originally read from the GIS map

Second, these patches are sorted by the development probability index, D. Next, locations are selected from this list to be developed. If neighborhoods are not being developed, the first n locations are chosen for development, where n is the user-selected res_per_year value. If the neighborhood switch is set to "On," then each selected patch recursively induces growth in its immediate neighbors until a new neighborhood is developed with a total size less than or equal to the user-set neighborhood_size variable. All newly developed areas then increase the urban attractiveness of neighboring patches as follows:

$$A = A + (1.0 - A) * N \tag{13.3}$$

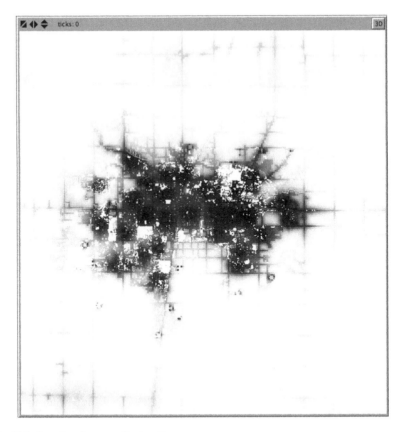

Fig. 13.2 Sample urban growth attractiveness map

where:

A = Base development probability index originally read from the GIS map
N = A user-settable neighbor-effect variable with a value between 0 and 1.0

Figure 13.2 shows the urban growth attractiveness map as displayed through NetLogo for the sample Champaign-Urbana area.

13.4 Simulation Experiments

This model provides a classroom exercise that allows the student to experiment with the various inputs: res_per_year, neighbor_effect, random_factor, neighborhood_size, and the on/off neighborhoods switch. Three "starter" experiments are described below.

13.4.1   Optimal Growth

Let us assume that the attractiveness input map accounts for all possible considerations that affect where people choose to develop. To do so, set the input values as
follows:

res_per_year	2000	This many 30×30 m patches will convert into urban patches each year
neighbor_effect	0	New urban patches will not induce any more adjacent new urban patches
random_factor	0	No randomness will be injected into the default attractiveness
neighborhoods	Off	Neighborhoods will not be developed
neighborhood_size	N/A	This value will not be used since *neighbors* is turned off
years	10	Run the model for 10 years

Hit one of the initialize buttons and then "run steps" to generate 10 years of
growth. If you repeat this experiment, the results will always be the same because
they are determined by the set values, with no randomization of attractiveness. Next,
reset the "random_factor" to 1.0, reinitialize the model, and run the simulation.
Note that urban growth throughout the area is now generated in a completely random
fashion, and that the random placements are different every time the simulation is
run. By setting the value somewhere between 0 and 1.0 you may seek an appropriate
balance where simulations are neither totally deterministic (i.e., determined solely
by the data inputs) nor totally random (i.e., completely overriding the set attractiveness values with random growth values).

13.4.2   Neighborhood Growth

You may find it disconcerting to see development happen one cell (or patch) at a
time if you are used to witnessing the development of 60 ha neighborhoods around
you, as is typical with suburban-style developments that are large and come into
existence all at once). To begin a neighborhood growth experiment, set the input
values as follows:

res_per_year	2000
neighbor_effect	0
random_factor	0.2
neighborhood_size	600
neighborhoods	Off
years	10

Initialize the model and run the designated steps. Note that the growth occurs
with some randomness, as indicated by the salt-and-pepper nature of the development
pattern, but is heavily guided by the urban growth attractiveness map. Now, switch

the neighborhoods value to "On" reinitialize and rerun. Note how the patch-level salt-and-pepper pattern has been replaced with clumps, which are themselves growing in random-shaped "blobs."

13.4.3 Induce Neighbor Growth

The level of urban growth attraction provided in the input map is, in part, based on proximity to already developed urban areas. Therefore, it is reasonable that all newly generated growth in the model should similarly affect the attraction of new growth in its neighbors. The neighbor_effect variable in the user interface provides a way to experiment with this concept. That variable identifies the difference, in terms of percentage, between a neighbor's growth attraction and 100% (a value of 1.0) that the attraction should be increased in response to next-door-neighbor development. To experiment with this variable, set the input values to those used in the previous experiment, specified in Sect. 13.4.2.

Initialize the model and run the designated steps. As before, the growth occurs with some salt-and-pepper randomness. Next, reset the neighbor_effect from 0 to 1.0, reinitialize the model, and rerun the simulation. Note that growth is now much more compact, with developed cells appearing more tightly clustered instead of spread-out, randomized patterns. This output represents what happens when new growth stimulates growth in neighboring cells.

13.5 Discussion

This model provides a starting point for exploring how new growth may induce growth in adjacent areas, how randomness impacts growth attractiveness, how differing sizes of new neighborhoods may lead to different results, and what effects different annual target levels of urban growth might have. Experimentation with this model will help a student to gain appreciation for the consequences of choosing different input variables, and it is useful for understanding the implications of different development styles on the landscape.

Driving forces behind urban growth attractiveness are represented in the form of the growth attractiveness input map, which can be enlarged or modified through GIS analyses for the sample locations or any other site for which appropriate GIS data are available.

13.6 Conclusions

This model provides students a more direct experience of basic planning theory and dynamics than is available through more passive learning processes alone. This experience can lead to more fruitful discussions and deeper insights, and may inspire further refinement of the model to increase its explanatory power. It is possible, for example, to extend the model to execute all of the GIS processing required to generate the urban attractiveness information based directly on raw transportation, land cover, elevation, and no-growth maps. One key to urban development is often the installation of utilities that are required to support residential areas; this model could be extended to capture those plans, thus improving its predictive capabilities. Master plans and zoning ordinances defining where residential development may or may not occur also could be captured in a modified version of this model. The model's built-in generic assumptions about residential development could be extended to distinguish between large-lot and dense apartment and condominium growth. The model also can provide a foundation for further exploration of ideas about urban growth through spatial simulation modeling.

References

Agarwal C, Green GM, Grove JM, Evans TP, Schweik CM (2002) A review and assessment of land-use change models: dynamics of space, time and human choice. US Forest Service, Northeastern Research Station, Burlington

Barredo JI, Kasanko M, McCormick N, Lavalle C (2003) Modelling dynamic spatial processes: simulation of urban future scenarios through cellular automata. Landsc Urban Plann 64:145–160

Calthorpe P, Fulton W (2001) The regional city. Island Press, Washington, DC

Clarke K, Gaydos LJ (1998) Loose coupling a cellular automaton model and GIS: long-term urban growth prediction for San Francisco and Washington/Baltimore. Int J Geogr Inf Sci 12:699–714

Costanza R, Voinov A (2001) Modeling ecological and economic systems with STELLA: part III. Ecol Model 143:1–7

Costanza R, Voinov A (2004) Introduction: spatially explicit landscape simulation models. In: Costanza R, Voinov A (eds) Landscape simulation modeling: a spatially explicit, dynamic approach. Springer, New York

Deal B, Schunk D (2004) Spatial dynamic modeling and urban land use transformation: a simulation approach to assessing the costs of urban sprawl. Ecol Econ 51:79–95

Echenique MH, Flowerdew AD, Hunt JD, Mayo TR, Skidmore IJ, Simmonds DC (1990) The MEPLAN models of Bilbao, Leeds and Dortmund. Transport Rev 10:309–322

EPA (2000) Projecting land-use change: a summary of models for assessing the effects of community growth and change on land-use patterns (EPA/600/R-00/098). U.S. Environmental Protection Agency, Office of Research and Development, Cincinnati

Ewing R, Pendall R, Chen D (2002) Measuring sprawl and its impacts. Smart Growth America, Washington, DC

Ford A (1999) Modeling the environment: an introduction to system dynamics modeling of environmental systems. Island Press, Washington, DC

Grimm V, Railsback SF (2005) Individual-based modeling and ecology. Princeton University Press, Princeton

Jantz CA, Goetz SJ, Shelley MK (2003) Using the SLEUTH urban growth model to simulate the impacts of future policy scenarios on urban land use in the Baltimore-Washington metropolitan area. Environ Plann B Plann Des 30:251–271

Klosterman R (1999) The what if? Collaborative planning support system. Environ Plann B 26(3):393–408

Landis J, Zhang M (1998) The second generation of the California urban futures model, part 1: model logic and theory. Environ Plann A 25(5):657–666

Putman SH (1983) Integrated urban models: policy analysis of transportation and land use. Routledge, Oxford

United Nations Population Fund (2007) State of world population 2007: unleashing the potential of urban growth. United Nations Population Fund, New York

Waddell P (2002) UrbanSim: modeling urban development for land use, transportation and environmental planning. J Am Plann Assoc 68(3):297–314

Waddell P, Ulfarsson GF (2004) Introduction to urban simulation: design and development of operational model. In: Haynes K, Stopher P, Button K, Hensher D (eds) Handbook 5: transport geography and spatial systems. Pergammon, London

Wang Y, Choi W, Deal BM (2005) Long-term impacts of land-use change on non-point source pollutant loads for the St. Louis metropolitan area, USA. Environ Manage 35(2):194

Westervelt, James, BenDor TK, Sexton JO (2011) A Technique for Rapidly Assessing Regional Scale Urban Growth. Environment and Planning B 38(1):61–81

Wilensky U (1999) NetLogo. Computer software. Northwestern University, Center for Connected Learning and Computer-Based Modeling, Evanston. http://ccl.northwestern.edu/netlogo/. Last accessed date 2/10/12

Chapter 14
Modeling Intimate Partner Violence and Support Systems

Marina Drigo, Charles R. Ehlschlaeger, and Elizabeth L. Sweet

14.1 Background

Intimate partner violence (IPV) is a serious ongoing social problem. IPV refers to the physical, psychological, emotional, and sexual abuse among intimate hetero-sexual partners (Hattery 2009). Both men and women can be victims of family abuse, but the percentage of affected women is higher (Catalano 2007; Hattery 2009). The National Violence Against Women Survey, conducted from November 1995 to May 1996, revealed that 22.1% of surveyed women and 7.4% of surveyed men have been physically assaulted by an intimate partner in her or his lifetime (Tjaden and Thoennes 2000). However, women appear to use violence primarily in self-defense and are more vulnerable to physical injuries than men (Hattery 2009; Tjaden and Thoennes 2000). IPV affects people of all races, income levels, and social classes, but African American, immigrant, and low-income women appear to be at a higher risk (Firestone et al. 2003; Garcia et al. 2005; Hattery 2009; Raj and Silverman 2002).

Due to such problems as ineffective data-collection systems, underreporting of incidents, and the use of different definitions of IPV by criminal justice, social services, and healthcare agencies (Hiselman 1999), there are ongoing debates about

M. Drigo (✉)
Department of Urban and Regional Planning, University of Illinois,
111 Temple Buell Hall, 611 Taft Drive, Champaign, IL 61820, USA
e-mail: marina.v.drigo@gmail.com

C.R. Ehlschlaeger
U.S. Army Engineer Research and Development Center, Construction Engineering
Research Laboratory, 2902 Newmark Drive, Champaign, IL 61822, USA

E.L. Sweet
Department of Geography and Urban Studies, Temple University,
308 Gladfelter Hall Temple University, Philadelphia, PA 19122, USA

J.D. Westervelt and G.L. Cohen (eds.), *Ecologist-Developed Spatially Explicit Dynamic Landscape Models*, Modeling Dynamic Systems, DOI 10.1007/978-1-4614-1257-1_14, © Springer Science+Business Media, LLC 2012

IPV rates. According to the US Bureau of Justice Statistics, the average annual rate of nonfatal victimization of females by an intimate partner is 4.2 per 1,000 persons of age 12 or over (Catalano 2007). Other estimates of annual IPV rates vary from 9.3 to 220 per 1,000 women, with the most commonly cited figure of 116 annual acts of any violence or 34 acts of severe violence per 1,000 women on the basis of the 1975 and 1985 National Family Violence Surveys (Crowell and Burgess 1996). IPV incidents rarely get reported to the authorities, and the share of incidents reported to police has been estimated to vary from 2 to 52% of the actual occurrences (Wolf et al. 2003).

There is a positive correlation between violence and the economic disadvantage of women (Basu and Famoye 2004; Farmer and Tiefenthaler 1997; Sanders and Schnabel 2006). Women from marginalized communities often make numerous attempts to leave an abusive relationship (Sullivan et al. 1992), but often cannot succeed because of the lack of access to employment, education, transportation, housing, child care, financial support, and legal support. This lack of structural support has been associated with short-term and long-term homelessness, housing instability (Bassuk et al. 2001; Baker et al. 2003), and the inability to find or maintain employment (Romero et al. 2003; Bell 2003).

Research indicates that a woman's cultural, ethnic, or social background can influence patterns of accessing and utilizing social services. Women have different frequencies of use and different perceived effectiveness of various help sources (Allen et al. 2004). According to Lipsky et al. (2006), non-Hispanic White women were nine times more likely to use emergency services and twice as likely to use domestic violence services when compared with Hispanic women. South Asian women were more likely to disclose abuse to family members or friends rather than official sources, but received a lower level of support from nonkin members when compared with Hispanics and African Americans (Yoshioka et al. 2003). Non-Hispanic women may be less likely to disclose abuse to a family member than Hispanics (Ingram 2007). Finally, the level of informational awareness regarding the availability of formal support can be as low as 45–50% among immigrant women (Raj and Silverman 2003; Murdaugh et al. 2004).

Usually IPV has been modeled using traditional statistical methods. To the best of our knowledge, this text documents the first project that attempts to represent IPV dynamics within a spatially explicit social support system. The model can help policy makers understand the dynamics and context of where IPV occurs, and it also can be used as a tool for testing policy and combating the aforementioned informational problems associated with IPV.

14.2 Objective

The objectives of this project are to model a body of knowledge developed by subject matter experts in order to represent IPV dynamics in the lives of women who do not respond equally to a one-size-fits-all set of remedies to violence at home.

While women from a higher socioeconomic context tend to solve such problems themselves (Hattery 2009), those from marginalized communities lack access to social services and other resources (Allen et al. 2004). We thus predict that better access to shelters, along with increased informational awareness of the public and cultural sensitivity of the service providers regarding IPV problems, should diminish IPV rates among women in general and women from lower socioeconomic strata in particular. It is intended that policy makers can use the model as a tool to evaluate the effectiveness of various IPV policies among major racial groups as well as insufficiently represented individuals without actually "experimenting" with people at risk. Accordingly, this model was designed to serve as a template that can be modified as needed in order to match the requirements of other users.

14.3   Model Description

14.3.1   Purpose

The purpose of the model is to evaluate the effectiveness that such parameters as cultural sensitivity of service providers, public awareness, and the number of shelter beds have on IPV rates among non-Hispanic White, African American, and Hispanic women, organized by income level, in Chicago, IL. The model demonstrates the discrepancy between officially reported violent incidents and all violent incidents, both reported and unreported. While many spatially explicit models represent the flow of a system across the duration of many generations of population or events, this model explores the state changes of a set of IPV events at a point in time until those IPV events are resolved. Figure 14.1 demonstrates the conceptual relationships in the model.

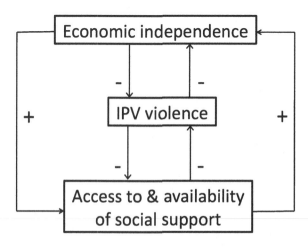

Fig. 14.1 Conceptual relationships

14.3.2   State Variables and Scales

The model was developed using Netlogo 4.1 (Wilensky 1999).[1] The model area is the City of Chicago represented by a map of census tracts mapped onto a 90-m grid. Each tract contains information on its racial composition to account for the percentage of African American, non-Hispanic White, and Hispanic individuals. A portion of Chicago's social-support system is represented by 60 community service centers that provide classes in English as a Second Language (ESL) and the General Equivalency Diploma (GED), child care, financial, and housing assistance. The model operates on 1-month time step. For the results reported here, the simulation was run for 10 years.

The model contains two types of agents: individual humans and shelters.

Representing the people in a study area is a major challenge in building models that need demographic input data, and that was the case during the development of our IPV model. At the time we developed the IPV model, the most recent available census data at the tract scale was based on 2009 estimates by GeoLytics, Inc. (2009). A free version of the 2009 estimates, obtained from the Joseph Regenstein Library at the University of Chicago, provided only the most basic attribute information, such as race, ethnicity, age, gender, and number of households by income per census tract. However, detailed household attribute information for 2008 was available from the American Community Survey, or ACS (US Census Bureau 2008). The ACS provides major attributes for people, including age, gender, race, ethnicity,[2] personal income, household income, type of household, educational attainment, school enrollment, English-speaking ability, number of children and children's age, vehicle, poverty status, relationship to the head of the household, public assistance (food stamps), and the year when a person moved to a household. However, ACS data do not have spatial resolution as detailed as census tracts. In order to estimate the location of relevant agents containing the necessary ACS variables, Digital Populations software (Ehlschlaeger 2004)[3] used land-use and census-tract spatial and proportional information to capture locations for the households and people (and their attributes) from the ACS. Digital Populations, at the time the IPV model was built, located the proportion of each race, ethnicity, age group, and income based on 2009 census tract spatial proportions of each attribute. Due to the lack of the proportions of marital status and English-speaking abilities by age in 2009 census tract data, these attributes were based instead on 2001 US census data. Households and person attributes were coregistered based on the 2008 ACS attributes. If we assume the 2008 ACS population attribute proportions are more accurate than the 2009 census tract proportions, then the population sample generated by Digital Populations underrepresents Hispanics by 2% and African Americans by 5%; and it overrepresents

---

[1] An operational copy of this model is available through http://extras.springer.com.

[2] The definitions of race and ethnicity used by the US Census Bureau may differ from the definitions used elsewhere in the literature. These variations account for slight differences between the model and certain standard references.

[3] Digital Populations is an open source software that may be downloaded for no cost at http://digitalpopulations.pbworks.com/w/page/26034597/FrontPage.

non-Hispanic Whites by 6%. The model's targeted sample of interest is drawn from 1% of Chicago's married or cohabiting African American, non-Hispanic White, and Hispanic women over the age of 15. Additional attributes are described in the Sect. 14.3.7, and the data for those variables were derived from the literature.

Shelters are the second element of the social support system. Following Levin et al. (2004) and the 2006 assessment of homeless population (Chicago Coalition for the Homeless 2006), it has been estimated that approximately 0.15% of homeless women are without homes due to IPV incidents, and 4.3% of all shelter beds are available for these women. The number of shelter beds in Chicago, including those in homeless and family shelters, is estimated to be 3,337, based on the data collected from the web sites of corresponding agencies. According to the Chicago Coalition for the Homeless, 73,656 people were homeless at some point in fiscal year 2006, which constitutes approximately 2.6% of Chicago's 2006 population (2,833,321 people as estimated by the US Census Bureau). Thus, the ratio of 3,337 beds to 73,656 people is 0.045 beds per person. The Chicago Coalition for the Homeless also estimated that adults constituted 60% of all homeless population, while women represented 43% of all homeless adults in 2004 (Library Index 2010), and 22% of women in shelters became homeless immediately following an IPV incident (Levin et al. 2004).

The value of 0.15% is a result of multiplying these percentages. Because this model uses a 1% sample of the population (30,537 people), it has been estimated that 45 women can be potentially homeless due to IPV incidents (i.e., 30,537 multiplied by 0.15) and the number of shelter beds is similarly reduced to 2 (i.e., 45 multiplied by 0.045). Children are not explicitly taken into account in the model, and each woman receives one bed regardless of the number of children she has.

The model uses two shelters with one bed in each. One shelter is located in a Hispanic tract and another one in a racially mixed tract. The choice of these locations was random.

14.3.3   Process Overview and Scheduling

Each sampled woman may experience violence at any time step. It is assumed that each battered woman wants to reduce or stop violence by becoming economically independent (i.e., having a higher income) or by leaving the relationship if the violence continues. Once a woman experiences violence, she can respond in any or combination of the following: call police, disclose abuse to friends (disclosure of abuse is assumed to be equal to asking for help), leave the relationship, go to a shelter or a community center, and become homeless or return back to the relationship. Figure 14.2 outlines general flows and state updates based on these activities.

Women living at or below the poverty level will search for a job and are considered capable of living apart from their abuser when their income reaches $16,640,[4] which

---

[4] This income is calculated on the basis of a 40-h week at $8.00 per hour, the Illinois minimum wage in 2010 (US Department of Labor, http://www.dol.gov/whd/state/stateMinWageHis.html). From the model's 1% sample of ACS households, all women at or below the poverty level have income lower than $16,000, with half of them earning less than $3,500.

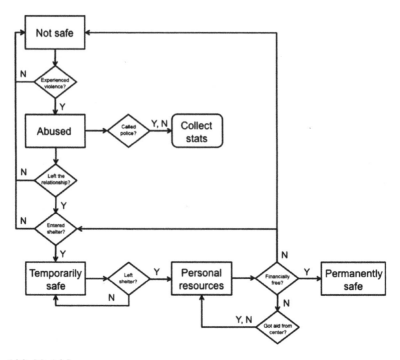

Fig. 14.2 Model flow

is 150% of the poverty level for a one-person household in 2009–2010 (US Census Bureau 2010). The job search is not directly triggered by exposure to violence, but the probability of becoming employed is influenced by violence. If a woman has not found a job and she has low educational attainment, poor English skills, or needs child care, she may go to a community center in search of these services.

Other processes include the willingness of a woman's friends to assist in response for her request for help and the provision of services by shelters.

14.3.4 Design Concepts

14.3.4.1 Adaptation

It is assumed that each battered woman's goal is to leave the relationship. As income is inversely related to the probability of violence, women at or below the poverty level want to obtain a job and reach income of $16,640. Searching for education and/or child care in community centers is a direct adaptive trait to increase employability. In general, searching for employment and higher income is an indirect adaptive trait since it may either help a woman to decrease the violence or to become financially independent and leave the relationship.

14.3.4.2   Sensing

It is assumed that women know about shelters and their location, but they do not know which shelter has available beds. It is also assumed that women know about community centers, but have no information about specific services. Women do not keep information about shelters and community centers that they visited in memory.

14.3.4.3   Interaction

Interactions between abused women and their friends occur only if the women have experienced violence. Women can ask friends for help, and friends may or may not provide assistance. Women do not keep track of friends who refused to help them. The number of friends and the specific individuals stay constant.

Women also interact with shelters and community centers by requesting services. Indirectly, they interact with other battered women which whom they compete for a shelter bed.

14.3.4.4   Stochasticity

The model is primarily stochastic. It uses stochasticity:

- As an input value with a mean and a standard deviation to the logistic regression equations in order to represent the variability of reality.
- As a decision-making process by comparing the results of a random number generator and the logistic regression equations in order to represent the uncertainty in the women's decisions.

Stochasticity is a vital characteristic to the model due to the unpredictability of human behavior in general and due to the variety of decisions that might be made in the uncertainty of the IPV context in particular.

14.3.4.5   Observation

Model results are collected based on the number of incidents and police calls (represented as a rate per 1,000 of the subject population) and the percentage of battered women who left the relationship categorized by race/ethnicity and income level (i.e., above and below poverty level).

14.3.5   Initialization

The model starts by initializing the geographic space. Next, shelters and community centers are created, followed by initialization of a 1% sample of Chicago's population.

Each married or cohabiting African American, non-Hispanic White, and Hispanic woman over the age of 15 is assigned an "at risk of domestic violence" status. None of the women are assigned any personal exposure to violence, and the shelters are empty. As the model runs, it takes about five time steps to fully populate the shelters.

The variables are read from the input files or created by using random numbers. The model does not use a seed, so the same parameters will produce different results for each simulation. At the end of the initialization, the racial composition of the tracts is calculated by summing up all individuals within the tract boundary and dividing it by the total population.

14.3.6 Input

The model's inputs include an ESRI shapefile to determine Chicago's boundary, a raster file of census tracts with a 90 m resolution, and population data produced using Digital Populations software (Ehlschlaeger 2004) based on the 2008 ACS. Each person is placed in geographic space according to the coordinates generated by Digital Populations and contains the variables discussed in Sect. 14.3.2.

For the data obtained from the literature, the values are assigned using a random number generator. Percentages are treated as probabilities. For example, if 5% of non-Hispanic White females over the age of 45 have alcohol problems, then each simulated non-Hispanic White female over 45 years old has that same probability of an alcohol problem.

Data on the location and available services in shelters and community centers are collected from the web sites of corresponding agencies.

14.3.7 Submodels

14.3.7.1 Experiencing Violence

The probability of violence in a couple is calculated based on a logistic regression developed by Salari and Baldwin (2002). They distinguish between verbal, physical, and injurious violence, but our model omits verbal violence. The probability of violence depends on variables such as household income, woman's income, presence of a child under 5 years old, official marriage status (vs. cohabiting), relationship duration, number of shared friends between a couple, race/ethnicity, number of personal friends, adherence to traditional gender roles, availability of informal support in a crisis, alcohol problems, and a level of self-esteem for each partner. These variables are either drawn from ACS information, or they are stochastically derived based on research on those social variables. If violence occurs, it is assumed to be equal to one act. "Experiencing violence" is one of the inputs into the *Calling police*, *Receiving help from friends*, *Leaving the relationship*, and *Looking for a job* submodels.

14.3.7.2 Calling Police

This submodel is initiated by exposure to violence and is based on a logistic regression developed by Felson and Paré (2005). It depends on variables including the reporting person's age, education, income, race/ethnicity, and whether the violence was initiated by a partner (vs. a stranger), whether it occurred at home (vs. on a street), and whether it resulted in a physical injury. It is assumed that IPV happens at home. Calls made to the police are documented for each woman on a yearly basis and represent official statistics reported to the authorities.

14.3.7.3 Disclosing Abuse to Friends

This submodel represents a request for any help, and is initiated for each woman that experienced violence during the simulation. Using a nonrepresentative sample, Yoshioka et al. (2003) and West et al. (1998) reported the percentages of women by race and ethnicity who are willing to discuss abuse with friends, with the model stochastically determining which abused women do so. A friend to be contacted is chosen randomly.

14.3.7.4 Receiving Help from Friends

The submodel is initiated if a woman contacted one of the friends and is based on a logistic regression developed by Beeble et al. (2008). It evaluates the willingness of an individual to provide any kind of support and depends on such variables as a person's gender, age, attitudes and beliefs about violence, perceived prevalence of violence rates in the community, and childhood and personal exposure to violence. The result is one of the inputs into *Experiencing violence* and *Leaving the relationship*. Public awareness (PA) is one of the dependent variables, and includes the level of willingness to disclose abuse to friends and the kinds of attitudes and beliefs about violence and perceived prevalence of violence rates in the community.

14.3.7.5 Leaving the Relationship

This submodel is initiated for each woman who has ever experienced violence and is based on a logistic regression developed by Sabina and Tindale (2008). It depends on variables including a yearly number of violent acts, the level of the most severe incident, harassment, power and control, health and depression, education, income, availability of social support, and type of employment.

14.3.7.6 Being Homeless

A woman's status is updated to homeless if she left the relationship and her income is below $16,640. During the homelessness period, a woman attempts to stay at a shelter (a temporary solution) or goes to a community center for a housing assistance (a permanent solution). A woman is allowed to stay homeless for up to 1 year. If she does not find housing assistance or a job during this period she returns to the abusive relationship.

14.3.7.7 Searching for Shelters

This submodel is initiated in two cases: when a woman leaves the relationship and when she becomes homeless. First, the probability of making a decision to go to a shelter depends on a woman's income, and is assumed to be 0.6 for women below the poverty level and 0.4 for women above the poverty level.[5] The probability of making an actual visit depends on physical distance, race/ethnicity, and the racial composition of the tract where a shelter is located. The formula for making an actual visit takes a form of:

$$P_{mk} = P_{CS} * 0.525e^{-0.11D_{ij}}$$

where P_{mk} is the percentage of trips for mode m and purpose k, P_{CS} is cultural sensitivity (CS) of service providers at shelter j, and D_{ij} is the distance in kilometers between a woman i and a shelter j. The parameters are fitted to automobile trips to health care clinics on the basis of Minneapolis data (Iacono et al. 2008) and, for the lack of better alternatives, are assumed to be representative of shelter visits.

The concept of cultural sensitivity of service providers is derived on the basis of Donnelly et al. (2005), who suggested that minority women may be less likely to use mainstream, white, middle-class oriented services that are usually located outside of their neighborhoods. The model's assumption is that if the racial composition of a tract where a shelter is located does not match a woman's race, then she will be less willing to use this service. The racial/ethnic composition of each census tract is defined on the basis of Turner and Hayes (1997): predominantly Black tracts are those where more than 50% of the population is African American; predominantly White tracts are those with less than 10% of African Americans and less than 10% of Hispanics; mixed tracts are those where African American population is between 10 and 50%; and Hispanic tracts are those with less than 10% of African Americans and more than 10% of Hispanics.

A woman gets accepted if a shelter has available beds. Otherwise, she moves to the next shelter. Women at or below the poverty level are allowed to stay at a shelter

---

[5] These values are informed by George et al. (2010), who reported significant differences in shelter utilization by employment status and number of hours worked, which implies differences by income level and poverty status. For simplification purposes, we used the poverty status divide rather than the actual employment status and number of hours worked.

for a maximum available time. The length of stay for women above the poverty level is determined by drawing a random number between 1 month and the maximum available time.

14.3.7.8 Looking for a Job

This submodel is initiated for each woman who is at or below 100% of the federal poverty level and is based on a logistic regression developed by Blumenberg (2002). It depends on such variables as education, English ability, ownership of a personal car, health problems (assumed to be equal to exposure to violence), and presence of children who need child care.

14.3.7.9 Searching for Services in Community Centers

This submodel is initiated in two cases: (1) when a woman is homeless and searches for housing assistance and (2) when she did not find a job and searches for GED or ESL classes or child care. A woman selects a community center within a distance of 3 km and evaluates the probability of visiting according to:

$$P_{ij} = \frac{A_j D_{ij}^{-\beta}}{\sum_{j=1}^{n} A_j D_{ij}^{-\beta}}$$

where A_j is the attractiveness o f a community center, D_{ij} is the distance in kilometers, and β is a coefficient equal to 1. Attractiveness of a community center is calculated as:

$$A_j = \sum WS$$

where W is a service and S is a coefficient representing perceived usefulness of a given service (Allen et al. 2004). Each service (i.e., education, housing assistance, child care, and financial aid) is coded as 1 if present and 0 if absent.

If a woman has decided to visit a community center, she does so, and checks for the services of interest. If no services are available, she selects the next block of community centers within the search distance.

14.4 Simulation Experiments

A total of 100 simulations were run to test the effect of the number of shelter beds at 100 and 200% increase, and the level of CS and PA at 50 and 100% increase in comparison with existing conditions (represented as 0% increase in the subsequent tables). The results were averaged for each experiment over 10 years of simulations.

Table 14.1 Average "real"/"official" incident rate per 1,000 women

Group	Total	Below poverty	Above poverty
Non-Hispanic White	6.1/1.1	34.2/6.0	5.3/1.0
Hispanic	19.1/3.5	56.0/10.2	11.6/2.0
African American	15.6/3.7	54.2/12.4	11.7/2.9

Table 14.1 demonstrates that the average "real" (as opposed to "official" incidents—those reported to the police) incident rate for a total population is the highest for Hispanic women, followed by African American and non-Hispanic White women. The model reveals noticeable differences when income is taken into account. Non-Hispanic White women experience the lowest incident rates in both the below and above the poverty level groups, while the rates for African American and Hispanic women in both groups are approximately the same. Non-Hispanic White women below the poverty level appear to experience violence 6.4 times more often than those from the above the poverty-level group. For Hispanic women below the poverty level, the incidence is 4.8 times higher than the more affluent group, and for African Americans the incidence is 4.6 times higher.

The number of incidents reported to police vs. actual number of incidents is substantially lower for each racial and income group. On average, non-Hispanic White women report 18.7% of actual incidents to police; Hispanic women report 18.3% of all incidents; and African American women report 23.7%. There are some differences by income group. Non-Hispanic White women below poverty report 17.5% of all incidents and those above poverty report 18.9%. For Hispanic women, the values are 18.2 and 17.2%, and for African American women the values are 22.9 and 24.8%, respectively.

The average percentage of women who were tagged as "safe" at the end of each simulation, and hence were out of the system either temporarily or permanently, is similar for non-Hispanic White, African American, and Hispanic women in the above the poverty level group (87.1, 86.8, and 86.0%, respectively). These values contrast significantly with the results obtained for women in the below the poverty level group: 60.8, 63.5, and 50.1% for White, African American, and Hispanic women, respectively.

Table 14.2 demonstrates the results for each change in parameters, and there is a great degree of variability in the results. This may be due to either a degree of uncertainty represented in the model, an inadequate population sample, or an insufficient number of simulation runs.

Linear regression analysis was carried out in order to determine the positive or negative effect of each parameter on simulation results. Linear regression indicates very low R^2 values and gentle slopes, as shown in Table 14.3. This table also shows variability in the signs of trends among the three racial groups. The same level of variability is observed for incident and safety rates, race, and income although the trends are slightly stronger when safety is used as a measuring parameter. Finally, the signs of the trends for incident and safety rates do not always match. Thus, it is

Table 14.2 Mean, standard deviation and standard error by simulation, total population[a]

Beds	PA increase	CS increase	Mean		
			White	African American	Hispanic
2	0	0	6.13	15.60	19.34
2	0	50	6.11	15.74	19.28
2	0	100	6.10	15.56	19.08
2	50	0	6.13	15.68	19.05
2	50	50	6.07	15.69	19.16
2	50	100	6.16	15.56	19.15
2	100	0	6.09	15.63	19.32
2	100	50	6.10	15.62	19.01
2	100	100	6.04	15.57	19.18
4	0	0	6.13	15.77	19.05
4	0	50	6.12	15.53	19.10
4	0	100	6.05	15.57	19.19
4	50	0	6.08	15.79	18.92
4	50	50	6.12	15.70	19.24
4	50	100	6.00	15.60	19.29
4	100	0	6.05	15.65	19.26
4	100	50	6.20	15.54	19.09
4	100	100	6.15	15.61	19.10
6	0	0	6.15	15.66	19.08
6	0	50	6.11	15.56	19.21
6	0	100	6.13	15.34	19.30
6	50	0	6.09	15.68	19.14
6	50	100	6.11	15.49	19.05
6	100	0	6.23	15.32	19.18
6	100	50	6.13	15.67	19.07
6	100	100	6.05	15.55	19.20

[a]Non-Hispanic White population had standard deviation from 0.36 to 0.49 and standard error from 0.04 to 0.05; African American had standard deviation from 0.88 to 1.20 and standard error from 0.09 to 0.12; Hispanic population had standard deviation from 0.87 to 1.24 and standard error from 0.09 to 0.12

expected that if a given scenario indicates an increase in incident rates, the safety rates should decrease and the converse should also be true. In several cases, however, a mismatch occurred instead.

14.5   Discussion

We expected that better accessibility to services together with informational and cultural awareness of the public and service providers would reduce the risk of IPV for all women, and those from marginalized communities in particular. However, the results of the simulations did not fully support our hypothesis.

Table 14.3 Simulation results for annual incident rates, total population

Grouped by	Scenario		White Slope	R^2	African American Slope	R^2	Hispanic Slope	R^2
Increased #	PA CS							
of shelter beds	0	0	0.0058	0.0005	0.0167	0.0007	−0.0642	0.0093
	0	50	−0.0001	4E-07	−0.0456	0.0052	−0.0174	0.0007
	0	100	0.0083	0.0009	−0.0542	0.0069	0.0552	0.0065
	50	0	−0.0099	0.0015	0.0005	6E-07	0.0224	0.001
	50	50	0.01	0.0014	−0.0495	0.0061	−0.028	0.0021
	50	100	0.0176	0.0036	−0.062	0.0081	0.0065	1E-04
	100	0	0.0116	0.0018	0.0113	0.0003	−0.0631	0.0088
	100	50	−0.0116	0.0019	−0.017	0.0008	0.0475	0.0048
	100	100	−0.0063	0.0005	−0.0197	0.0011	−0.0367	0.0029
Increased CS	Beds PA							
	2	0	−0.0003	0.0008	−0.0004	0.0002	−0.0026	0.0096
	2	50	0.0002	0.0004	−0.0011	0.0019	0.001	0.0016
	2	100	−0.0005	0.0018	−0.0006	0.0006	−0.0014	0.0024
	4	0	−0.0008	0.0062	−0.0021	0.0068	0.0014	0.0031
	4	50	−0.0008	0.0048	−0.0019	0.0051	0.0037	0.0203
	4	100	0.001	0.0083	−0.0004	0.0002	−0.0017	0.0039
	6	0	−0.0002	0.0005	−0.0032	0.0187	0.0022	0.006
	6	50	−0.0002	0.0003	0.0012	0.0023	−0.0007	0.0006
	6	100	−0.0012	0.0118	−0.0018	0.0064	−0.0003	0.0002
Increased PA	Beds CS							
	2	0	−0.0004	0.0015	0.0003	0.0002	−0.0002	4E-05
	2	50	−0.0001	0.0002	−0.0013	0.0025	−0.0027	0.0112
	2	100	−0.0006	0.0023	0.0001	1E-05	0.001	0.0014
	4	0	−0.0008	0.0058	−0.0013	0.0024	0.0022	0.0069
	4	50	0.0008	0.0066	8E-05	1E-05	−7E-05	7E-06
	4	100	0.001	0.0072	0.0004	0.0003	−0.0009	0.0014
	6	0	−0.0002	0.0003	1E-04	2E-05	−0.0001	2E-05
	6	50	−0.0006	0.0034	−0.0001	2E-05	−9E-05	1E-05
	6	100	−0.0011	0.0105	0.0015	0.0039	−0.0026	0.0089

According to most studies, the rate of reported IPV is higher for African American women than Hispanics, and the rates for non-Hispanic White and Hispanic women do not significantly differ from one another (Tjaden and Thoennes 2000). However, as noted in the introduction, only 2–52% of actual IPV incidents are reported to the authorities. The model predicts the highest rates of incidents actually occurring for Hispanic women. These simulation results agree with studies that identify an inverse relationship between income and IPV rates (Catalano 2007): the sampled Hispanic women had a mean personal income of $17,746, while African Americans had a mean income of $29,507 and non-Hispanic White had a mean personal income of $39,311.

The simulations indicate that different scenarios produce different trends in the average incident rates and the percentage of women in safety, as sorted by race and income. In some simulation scenarios a higher percentage of women left the system than in others, yet this dynamic was not reflected in the average incident rates. As such, the main question of interest is why a given combination of parameters would potentially increase violence rates for one group of women while decreasing violence rates for another group.

Women from the various socioeconomic and racial groups have very different levels of exposure to violence, and the ability of low-income women in the modeled system to leave IPV relationships appears to be substantially lower than for others. One possible explanation for why none of the simulated scenarios demonstrates a definitive impact on average annual IPV rates is that higher-income women with less IPV experience leave the IPV relationship faster, and thus reach safety in higher percentages than lower-income women with more IPV experience.

One result we expected was that high public awareness of IPV would diminish annual IPV rates by increasing the probability of leaving the relationship. Benefiting from increased public awareness, more women with higher income may leave an abusive relationship and enter a shelter. However, women with lower income tend not to leave the relationship, and if they do, then they compete for shelter space with the rest of the population. Because shelters in the model select their future residents at random, low-income women had a lower chance of selection simply because they are underrepresented in the total population of those requesting shelter beds. For this reason, increasing the number of shelter beds in the model may not show statistically significant patterns of change because the shelters do not have a realistic opportunity to work with this particular segment of the population in need.

If more beds were added to the model, the simulations might show a significant reduction in IPV rates. However, a shelter is only a temporary solution both in the model and in the real world. In the model, specifically, a shelter serves as a temporary holder for a given number of time steps, during which a shelter resident is considered safe. As such, it is possible that the model would demonstrate stronger trends if unlimited shelter beds were available. However, without more support in the areas of housing, employment, and education, low-income women would simply cycle between shelters until the end of the simulation, remaining safe as the result of a scenario that is not actually sustainable in the real world.

Another result we expected was that a higher level of cultural sensitivity by service providers would have a positive impact on shelter visits and safety rates among non-Hispanic White and African American women, as those women would be more willing to visit shelters located in tracts whose racial composition does not match their own. Because one of the shelters is located in a Hispanic neighborhood and another in a racially mixed neighborhood, a high level of cultural sensitivity would not be expected to have the same effect among Hispanic women in the model. However, the simulation results unexpectedly revealed a negative impact on shelter visits and safety rates among Hispanic women. It is possible that as more non-Hispanic White and African American women start requesting and receiving shelter beds, a lower percentage of Hispanic women will enter shelters because there is less room for them than previously. In other words, the model suggests that as Hispanic women are exposed to a higher rate of IPV, greater cultural sensitivity at shelters may in some circumstances increase the total violence rates because the population most at risk is less able to obtain shelter due to competition with other groups.

The location of shelters is also important. If the location is not geographically convenient for most of the population in need, then at some point higher numbers of shelter beds and higher public awareness or cultural sensitivity may cease to have a positive impact because women are less likely to visit. Policy makers may need to consider whether it is more effective to have one large shelter with many beds or many small shelters distributed more widely. Policy makers also need to consider how shelter location may affect the probability of visits by women from different socioeconomic groups. A trend that results from adding more shelter beds and making changes in public awareness and cultural sensitivity may have either positive or negative implications, depending on the demographics of the affected population. For example, if more low-income women obtain a shelter bed, the trend may indicate a reduction in annual IPV rates; but if higher-income women occupy the shelter beds, the trend may indicate an increase in IPV rates.

The frequency of the calls to the police as sorted by race and income was not addressed by our hypothesis. Police calls represent the "official" incident rates and are used for purposes of model validation. The low number of police calls relative to the number of actual IPV incidents in the model is in agreement with the literature (Wolf et al. 2003). Additionally, African American women in the model report to police more often than non-Hispanic White women, which also correlates to the literature (Catalano 2007). The simulated rate of police reports by African American women in the model is slightly higher than the rate by Hispanic women; this, too, correlates to the literature (Catalano 2007), but additional research is required to determine the statistical significance of this difference. In contrast to some studies (Felson and Paré 2005; Pearlman et al. 2003), the simulation results indicate that higher-income non-Hispanic White and African American women appear to report to the police slightly more often than lower-income women. Similarly, additional research is needed to determine whether the difference is statistically significant.

14.6   Conclusions

In this chapter we have presented the framework for a new application of agent-based modeling to capture the dynamics of IPV. Building on socioeconomic and cultural representations of IPV, the model simulates the help-seeking behavior of battered women within informal and formal social support systems: friends, shelters, and community service centers. It demonstrates a novel method of representing and understanding the dynamics of IPV, and can be used as a tool for testing the implications of various policy alternatives.

This model should be considered a prototype, and as such it could benefit significantly through further development. Four immediate adjustments would enhance the model's ability to address the great variety of ways in which women respond to IPV. First, household and population input data generated by Digital Populations would have benefited if additional attributes, such as marital status,

English language abilities, and poverty status, were available for the 2009 census tract data. Second, the version documented here assumes a negative feedback between violence and economic independence (i.e., greater economic resources reduce exposure to violence while fewer resources increase exposure to violence). In our model, economic independence is linked to leaving the abusive relationship. However, this connection may not hold true for all women. Adjustments should account for a possible backlash effect in which the chance of IPV increases as a woman gains more economic power in a way that challenges the man's established authority in the relationship. Third, the model could benefit from further development that enables it to more accurately represent the decisions and opportunities for women who do not have a viable path for leaving the relationship. Finally, the current version of the model only represents married and cohabiting households as defined by the US Census Bureau. It should be further refined to account for divorced and separated individuals, who appear to experience IPV at a much higher rate (Catalano 2007). In addition to these four improvements, this model also could be extended to help better inform policy decisions about where to locate future shelters and community service centers. These improvements of the model would help policy makers to better understand how their decisions may affect different populations.

Acknowledgment This research was partially supported by the US Army Engineer Research and Development Center (ERDC) through a Center-Directed Research project entitled "Rapid Model Prototyping for Infrastructure and Essential Services." The authors express their appreciation to Dr. Eric Dimperio of the ERDC Environmental Laboratory, Vicksburg, MS, for his assistance in cleaning up the NetLogo code.

References

Allen NE, Bybee DI, Sullivan CM (2004) Battered women's multitude of needs: evidence supporting the needs for comprehensive advocacy. Violence Against Women 10(9):1015–1035

Baker CK, Cook SL, Norris FH (2003) Domestic violence and housing problems: a contextual analysis of women's help-seeking, received informal support, and formal system response. Violence Against Women 9(7):754–783

Bassuk EL, Perloff JN, Dawson R (2001) Multiply homeless families: the insidious impact of violence. Hous Policy Debate 12(2):299–320

Basu B, Famoye F (2004) Domestic violence against women, and their economic dependence: a count data analysis. Rev Polit Econ 16(4):457–472

Beeble ML, Post LA, Bybee D, Sullivan CM (2008) Factors related to willingness to help survivors of intimate partner violence. J Interpers Violence 23(12):1713–1729

Bell H (2003) Cycles within cycles: domestic violence, welfare, and low-wage work. Violence Against Women 9(10):1245–1262

Blumenberg E (2002) On the way to work: welfare participants and barriers to employment. Econ Dev Q 16(4):314–325

Catalano S (2007) Intimate partner violence in the United States. US Department of Justice, Bureau of Justice Statistics, Washington, NCJ 210675

Chicago Coalition for the Homeless (2006) How many people are homeless in Chicago? An FY 2006 analysis. Survey Research Laboratory at the University of Illinois at Chicago, Chicago. http:\\www.chicagohomeless.org/files/images/Homelessestimate2006.doc. Accessed 01/2010

Crowell NA, Burgess AW (eds) (1996) Understanding violence against women. National Academy Press, Washington

Donnelly D, Cook K, van Ausdale D, Foley L (2005) White privilege, color blindness, and services to battered women. Violence Against Women 11(1):6–37

Ehlschlaeger CR (2004) Digital Populations: Building Multiple Realizations of Population for Cluster Detection Analysis. Accepted for Presentation at GIScience2004 Conference, College Park Campus, Maryland, USA, October 20, 2004. http://chuck.Ehlschlaeger.info/2004/giscience2004

Farmer A, Tiefenthaler J (1997) An economic analysis of domestic violence. Rev Soc Econ 55(3): 337–358

Felson RB, Paré P (2005) The reporting of domestic violence and sexual assault by nonstrangers to the police. US Department of Justice, National Institute of Justice, Washington, NCJ 209039

Firestone J, Harris RJ, Vega WA (2003) The impact of gender role ideology, male expectancies, and acculturation on wife abuse. Int J Law Psychiatry 26:549–564

Garcia L, Hurwitz E, Kraus J (2005) Acculturation and reported intimate partner violence among Latinas in Los Angeles. J Interpers Violence 20(5):569–590

GeoLytics, Inc (2009) Estimates standard 2009: 2009 estimates and 2014 projections. [Cd ROM]. GeoLytics, East Brunswick

George C, Grossman S, Lundy M, Rumpf C, Crabtree-Nelson S (2010) Analysis of shelter utilization by victims of domestic violence: quantitative analysis. Illinois Criminal Justice Information Authority, Grant # 06-DB-BX-0043

Hattery AJ (2009) Intimate partner violence. Rowman & Littlefield Publishers, Inc, Lanham

Hiselman J (1999) Intimate partner violence in Illinois. Illinois Criminal Justice Information Authority: Trends and Issues Update 1(8). Available at http://www.icjia.state.il.us/public/pdf/tiupdate/dv.pdf. Accessed 12/2009

Iacono M, Krizek K, El-Geneidy A (2008) Access to destination: how close is close enough? Estimating accurate distance decay functions for multiple modes and different purposes. Minnesota Department of Transportation, St. Paul, MN/RC 2008-11

Ingram E (2007) A comparison of help seeking between Latino and non-Latino victims of intimate partner violence. Violence Against Women 13(2):159–171

Levin R, Lise M, Jody R (2004) Pathways to and from Homelessness: Women and Children in Chicago Shelters. Available at http://www.impactresearch.org/publication/publicationdate.html. Accessed 8/30/2009

Library Index (2010) The demographics of homelessness—profiles of the homeless. http://www.libraryindex.com/pages/2287/Demographics-Homelessness-PROFILES-HOMELESS.html. Accessed 12/2010

Lipsky S, Caetano R, Field CA, Larkin GL (2006) The role of intimate partner violence, race, and ethnicity in help-seeking behaviors. Ethn Health 11(1):81–100

Murdaugh C, Hunt S, Sowell R, Santana I (2004) Domestic violence in Hispanics in the Southeastern United States: a survey and needs analysis. J Fam Violence 19(2):107–115

Pearlman D, Zierler S, Gjelsvik A, Verhoek-Oftedahl W (2003) Neighborhood environment, racial position, and risk of police-reported domestic violence: a contextual analysis. Public Health Rep 118(1):44–48

Raj A, Silverman J (2002) Violence against immigrant women: the roles of culture, context, and legal immigrant status on intimate partner violence. Violence Against Women 8(3):367–398

Raj A, Silverman J (2003) Immigrant South Asian women at greater risk for injury from intimate partner violence. Am J Public Health 93(3):435–437

Romero D, Chavkin W, Wise P, Smith L (2003) Low-income mothers' experience with poor health, hardship, work, and violence: implications for policy. Violence Against Women 9(10):1231–1244

Sabina C, Tindale RS (2008) Abuse characteristics and coping resources as predictors of problem-focused coping strategies among battered women. Violence Against Women 14(4):437–456

Salari SM, Baldwin BM (2002) Verbal, physical and injurious aggression among intimate couples over time. J Fam Issues 23(4):523–550

Sanders CK, Schnabel M (2006) Organizing for economic empowerment of battered women: women's savings accounts. J Community Pract 14(3):47–68

Sullivan C, Basta J, Tan C, Davidson WS II (1992) After the crisis: a needs assessment of women leaving a domestic violence shelter. Violence Vict 7(3):267–275

Tjaden P, Thoennes N (2000) Full report of the prevalence, incidence, and consequences of intimate partner violence against women: findings from the national violence against women survey. US Department of Justice, National Institute of Justice, Washington, NCJ 183781

Turner MA, Hayes C (1997) Poor people and poor neighborhoods in the Washington metropolitan area. http://www.urban.org/publications/407425.html. Accessed 02/2010

U.S. Census Bureau (2008) American Community Survey (ACS): Public Use Microdata Sample (PUMS). http://factfinder.census.gov/home/en/acs_pums_2008_1yr.html. Accessed 9/2009

U.S. Census Bureau (2010) Poverty thresholds. http://www.census.gov/hhes/www/poverty/data/threshld/index.html. Accessed 8/2010

West C, Kantor GK, Jasinski J (1998) Sociodemographic predictors and cultural barriers to help-seeking behavior by Latina and Anglo American battered women. Violence Vict 13(4):361–375

Wilensky U (1999) NetLogo. Computer software. Northwestern University, Center for Connected Learning and Computer-Based Modeling, Evanston. http://ccl.northwestern.edu/netlogo/ Accessed 08/2009

Wolf ME, Ly U, Hobart MA, Kernic MA (2003) Barriers to seeking police help for intimate partner violence. J Fam Violence 18(2):121–129

Yoshioka MR, Gilbert L, El-Bassel N, Baig-Amin M (2003) Social support and disclosure of abuser: comparing South Asian, African American, and Hispanic battered women. J Fam Violence 18(3):171–180

Index

A

Agent-based models (ABMs), 175, 176.
　　See also Striped newt
　　metapopulation dynamics
American Community Survey (ACS), 238

B

BehaviorSpace tool, 38

C

Cave crickets, 48
Cave stabilization, 48
Cellular automata (CA)
　　interactions, 13
　　models, 175, 176
Colony invasion process, 46
Compilation errors, 22–23
Computer simulation model, 1–2.
　　See also Red Imported
　　Fire Ant (RIFA)
Conceptual model development
　　ad hoc model, 3
　　cognitive, probability, 2
　　communication, 3
　　complex task learning, 3
　　computer-based simulation model, 1–2
　　decision-support tool, 5
　　ecological models, 1
　　encoding, 2
　　expert knowledge, 4, 5
　　geospatially explicit model, 2
　　iteration and verification method, 5
　　knowledge transfer and documentation, 4

　　mathematical models, 1
　　neural network training, 4
　　objectives, 1, 5–6
　　quantitative tools, 4
　　simulation models, 4, 5
　　statistical analysis, 1
　　"sub-model," 3
Constraints modeling
　　potential model components, 12
　　potential model interactions, 12–13
　　simulation timeframe, 14
　　spatial resolution options, 15, 16
　　time step options, 14–15

D

Data analysis, 39
Demographic sensitivity analysis
　　adult mortality rates, 126
　　baseline *vs.* maximum parameter
　　　　value, 127
　　egg-to-age 1 mortality, 126
　　IBM *vs.* PVA model, 122, 123
　　parameter values, 123
　　population sizes, 126
　　sensitivity, 125
Digital elevation model (DEM), 67
Discrete mobile objects, 12

E

Energy exchange network, 46
Event driven approach, 14–15
Expedient models, 45
Expert knowledge, 4, 5

J.D. Westervelt and G.L. Cohen (eds.), *Ecologist-Developed Spatially Explicit*
Dynamic Landscape Models, Modeling Dynamic Systems,
DOI 10.1007/978-1-4614-1257-1, © Springer Science+Business Media, LLC 2012

F
Feral hog population control methods.
        See Hunting and contraception
        evaluation
Fixed resolution method, 15
Flood-pulse concept (FPC), 152
Foraging range, 48–49
FRAGGLE model
    biodiversity, 171
    design concepts
        density-dependent approach, 183
        habitat carrying capacity, 181
        lineage array, 183, 184
        lineage mixing index, 184
        movement decision, 181
        subpopulation mixing, 181–183
    dispersal behavior
        base case scenario, 187, 188, 190
        land-use change scenario, 187, 189, 190
    dispersal spread map, 187
    fragmentation modeling
        agent-based models, 175, 176
        cellular automata models, 175, 176
        species-specific modeling, 174
        trans-matrix species, 174
    GAP, 175, 186, 187
    gopher tortoise
        habitat attractiveness and user-defined
            dispersal mortality rate, 180
        hatchlings, 179
        natural history, 172, 173
        STELLA diagram, 178, 179
    habitat and home range characteristics,
        172–173
    habitat fragmentation, definition, 171
    human-induced management, 191
    land management activities, 191
    land use and land cover patterns, 186
    LUC, 172, 176, 190
    model parameterization, 184–185
    NetLogo 4.1, 177
    policy implications, 192
    population viability analysis, 177
    spatial-dynamic model, 176
    state variables and scales, 178
    west-central Georgia, 173–175

G
Gap Analysis Program (GAP), 175, 186, 187
Geographic information system (GIS),
        215–216, 224
    capabilities, 39
    interactions, 13

Geospatially explicit model, 2
Gopher tortoise (*Gopherus polyphemus*)
    carrying capacity, 125
    charismatic SAR, 85
    demographic parameter, 124
    Fort Stewart Army installation, 110
    habitat conditions, 127, 128
    IBMs, 110
    management strategy, 86
    military training activity, 105
    model description, 87–89, 111, 112
        adaptation and fitness, 115
        aging, 120
        basal area, 117
        better habitat search, 118–119
        carrying capacity map, 117, 118
        design concepts, 91
        dispersal probability, 119
        emergence, 115
        eviction from patch, 118
        Fort Benning study area, 87, 88
        gray scales, 94, 95
        growth of vegetation, 97
        habitat class map, 118
        individual-based simulation model, 87
        individual (agent) variables, 113–114
        initialization, 91–92, 116–117
        management level, 112
        model interaction, 116
        model interface, 92, 93
        model simulation dynamics, 96–97
        mortality, 120
        NetLogo, 87, 111
        patch variables, 112, 113
        process eggs and hatchlings, 98
        process overview and scheduling,
            90–91, 114–115
        reproduction submodel, 119–120
        sensing, 115
        set tortoise carrying capacity, 98
        soils suitability map, 94
        spatial and temporal scale, 112–113
        spatial habitat, 93–94
        state variables and scales, 89, 90
        stochasticity, 116
        study site, 87, 89
        tortoise death, 101
        tortoise development, 95–96
        tortoise feeding, 100
        tortoise growth, 99, 100
        tortoise migration, 99
        tortoise movement, 98–99
        tortoise reproduction, 100–101
        upland habitat type, 94

user-interface plots, 116
vector maps, 117
vegetation density map, 94
Vortex, 111
objectives, 87, 111
optimistic value, 125
population models, 109
PVA, 86, 109–110
range, 85, 86
resource managers, 129
simulation experiments
   demographic sensitivity analysis,
      122–124
   land management decisions, 104
   model calibration, 120
   model validation, 120–121
   population trends and probability
      of extinction, 121, 122
   SimGT model, 102, 103
   simple circular habitat experiment,
      101–102
   woody vegetation, 102
   100-year simulations, 103, 104

H
Hedonic logic approach, 228
Heterotrophic carbon stocks, 156, 164, 166
   biomass, 158
   carbon lost, 159
   consumption, 158
   physiological loss, 159
   prey and space limitation functions,
      158–159
   stock-specific values
      prey limitation, 159
      space limitation, 159, 160
Hierarchical models, 15
HubNet tool, 28
Hunting and contraception evaluation
   agent-based modeling system
      NetLogo 4.0.2, 135
   average population, 145, 147
   behavioral adaptations, 148
   contraceptive bait intensity
      and no hunting, 145, 146
   design concepts, 137–138
   events, 137
   fauna and flora, 134
   feral swine control, 133, 135
   Fort Benning and feral pig population,
      135, 136
   hunt and bait levels, 144

   hunting intensity and no
      contraceptive bait, 145
   input, 138–139
   landscape characteristics, 138
   low contraceptive bait intensity, 145, 146
   low hunt intensity, 145, 147
   NetLogo interface, 144
   pre-control and control periods, 137
   state variables and scales, 135–136
   submodels
      attrition, 140–141
      control methods, 143
      diet, 141–142
      farrowing, 140
      habitat, range and travel, 142–143
      recruitment, 139
      social grouping and dynamics, 141
Hypothesis testing
   loss of entire caves, 52, 53, 56
   management, 52, 53
   marginal mean estimation, 55
   RIFA, 55
   sensitivity, 53
   surviving crickets, 54
   "worst" case scenario, 53

I
Individual-based model for metapopulations
      on patchy landscapes-genetics and
      demography (IMPL-GD), 206–209
   conservation resources, 197
   demographic mortality, 202
   design concepts
      adaptation, 200
      emergence, 199
      fitness and interaction, 200
      observation, 200–201
      sensing and stochasticity, 200
   dispersal behavior, 202–204
   habitat parcel elimination, 199, 202
   habitat qualities, 201
   landscape initialization and initial
      whatsit distribution, 202
   metapopulation extinction, 198, 203
   multilinear regression analysis, 206
   NetLogo 4.0.4, 198
   offspring generation, 202
   simulated metapopulation percentage,
      204, 205
   state variables and scales, 199
   statistical analysis, 198
Individual-based models (IBMs), 110

Intimate partner violence (IPV)
   average real/official incident rate, 246
   community center and shelters,
      interactions, 241
   conceptual relationships, 237
   cultural sensitivity, 249–250
   definition, 235
   Digital Populations, 250–251
   direct adaptive trait, 240
   ESRI shapefile, 242
   geographic space initialization, 241–242
   linear regression analysis, 246, 248
   mean, standard deviation and standard
      error, 246, 247
   model flow and state updates, 239, 240
   police vs. actual number of incidents, 246
   policy makers, 250
   race/ethnicity, 241
   risk reduction, 247
   shelters information, 241
   socioeconomic context, 237
   state variables and scales, 238–239
   stochasticity, 241
   submodels
      abused women, 243
      being homeless, 244
      calling police, 243
      community center services, 245
      federal poverty level, 245
      logistic regression, 242, 243
      public awareness, 243
      shelter searching, 244–245
   violence vs. economic disadvantage,
      women, 236

L
Landscape patches, 12
Land-use change (LUC), 172, 176, 190
Linear objects, 12

M
MathematicaLink, 40
Mississippi river
   Adaptive Hydraulics 2D hydraulic
      simulation model, 155
   autotrophic stocks, 156–158, 164–165
   built structures, 153
   carbon fixation, 152, 167
   decomposers, 165, 166
   detritus, POC and DOC, 161, 165, 166
   ecological and hydraulic process, 154
   floodplain-river ecosystem, 151
   food web, 152

   grayscale GIS maps, 155–156
   heterotrophic stocks, 156, 164, 166
      biomass, 158
      carbon lost, 159
      consumption, 158
      physiological loss, 159
      prey and space limitation functions,
         158–159
      prey limitation, stock-specific values,
         159
      space limitation, stock-specific values,
         159, 160
   hydrology input change, 155
   influxes, 162, 163
   maps division, 157
   natural disturbance impacts, 168
   NetLogo 4.0.4, 155
   NetLogo grayscale map (see NetLogo,
      grayscale map)
   organic carbon, 151–152
   organic suspended sediments data, 157
   Pool 5 map, 153, 154
   preliminary model validation, 166–167
   productivity patterns, 152
   short-term and long-term monitoring
      programs, 153–154
   starting values, carbon stocks, 162, 163
   trigonometric design, 162
Mobile object interactions, 13
Model validation
   cave crickets, management process, 51
   cave sizes, 51
   raiding caves, 52
   RIFA, reliability, 50
   sensitivity effect, 51
Mounds
   foraging range, 49
   management, 50
   propagation and raiding, 49
Multidisciplinary group modeling projects
   constraints modeling
      potential model components, 12
      potential model interactions, 12–13
      simulation timeframe, 14
      spatial resolution options, 15, 16
      time step options, 14–15
   full dummy model construction, 20
   full model conceptualization, 16–17
   group dynamics
      leadership, 11–12
      model development and integration
         responsibility, 10–11
      scheduling, 11
      subcomponent development
         efforts, 11

model-development process, 7–8
model dissemination, 24
model integration
    demonstration, end users, 23–24
    logic error debug, 23
    NetLogo compilation error debug,
        22–23
objectives and scope identification, 8–9
resources identification
    computer technology, 10
    data availability, 9–10
    participant availability, 9
    personnel capabilities, 9
submodel construction, 21, 22
submodel design
    identification, 18
    model identification requirements, 20
    set submodel requirements, 18–20
three-stage process, 7

N
National Land Cover Data (NLCD) map,
        217, 225, 227–229
NetLogo. *See also* Pathway analysis
        through habitat (PATH)
        algorithm
    Champaign-Urbana area, 228, 230
    grayscale map
        autotrophic stocks, 163, 164
        decomposer stocks, 163, 165
        DOC, POC and detritus, 163, 165
        heterotrophic consumer stocks,
            163, 164
        regional urban growth modeling, 224
NetLogo modeling environment
    capabilities and features, 28–29
    coding
        accessibility, 36
        breeds, 36, 37
        construction, primitives, 36
        NetLogo language, 35
        user-defined procedure, 37
        Wolf Sheep Predation model, 34, 35
    compilation and execution processes, 38
    data input and output capabilities, 38
    extensibility, 39, 40
    information tab, 32, 33
    initialization process, 37–38
    installation and setup, 29
    interface tab
        buttons and widgets, 31
        Command Center, 30–32
        observer and *link* agent, 31

ticks, 30
turtles and *patches*, 31
    Wolf Sheep Predation model, 29–31
procedures tab, 32–34
program overview, 27–28
simulation experiments, 38
syntax and coding errors, 38
workflow description, 35
NetLogo-R extension, 39
NLCD map. *See* National Land Cover Data map

P
PATH algorithm. *See* Pathway analysis
        through habitat algorithm
Pathway analysis through habitat (PATH)
        algorithm
    animal movement choices and behaviors, 220
    design concepts, 214–215
    GIS, 215–216
    gopher tortoises, 216
    habitat connectivity, 211
    initialization, 215
    least cost method, 220
    location-specific maps, 214
    low overhead modeling approach, 220
    metapopulation, definition, 212
    migration density, 218
    net in-and out-migration, 219
    NLCD maps, 217
    patch lethality, 214
    radio tracking and telemetry studies, 221
    species migration, 213
    state variables and scales, 213
    submodels, 216
    walkers (virtual animals), 212
    wall-clock time, 218
Population logistic growth curve, 102
Population viability analysis (PVA), 86
Predator-prey relationships, 46

R
Red Imported Fire Ant (RIFA)
    aggressive omnivores, 44
    average number of crickets, 57, 58
    and cave crickets, 44
    cave cricket survivorship, 56
    cave loss prediction, 56–57
    colonization, 59
    disadvantages, 59
    field research, 58
    key gaps, 60
    loss of crickets, 58

Red Imported Fire Ant (RIFA) (*cont.*)
  management strategy, 44
  model description
    design concepts, 46, 47
    initialization and input, 47, 48
    potential impact, 45
    state variables and scales, 45, 46
    submodels, 48–50
  objectives, 44–45
  populations, 59
  raiding, 56
  recommendations, 61
  resource limitations, 59
  sensitivity, 58
  simulation experiments
    hypothesis testing (*see* Hypothesis
      testing)
    main and combined effect parameters, 50
    model validation, 50–52
    Tukey-Kramer procedure, 50
Regional urban growth modeling
  binary logistic regression model, 229
  classroom-level modeling exercise,
    225, 230
  computer-based simulation model, 223
  design concepts, 227
  events, 227
  hedonic logic approach, 228
  human decision-making, 223
  induce neighbor growth, 232
  initialization, 227
  land use change, 224
  model interface, 225, 226
  model logic, 229–230
  neighborhood growth experiment, 231–232
  NetLogo, 224
  NLCD map, 227–229
  optimal growth, 231
  state variables and scales, 225–226
  urban simulation models, 224
ReLogo, 40
Resources identification
  computer technology, 10
  data availability, 9–10
  participant availability, 9
  personnel capabilities, 9
River continuum concept (RCC), 152
Riverine productivity model (RPM), 152

S
Scala langauge, 27
SimGT model, 102, 103
Species at risk (SAR), 85
StarLogo, 27
Striped newt metapopulation dynamics
  breeding, 76
  climatic and habitat variables, 66
  correlation coefficient, 78–81
  DEM, 67
  design concepts, 70–71
  Ellabelle loamy sand areas, 167
  input, 72–73
  long-term metapopulation, 66
  model calibration, 76–78
  model process sequence, 69
  mortality, 75–76
  movement types, 74–75
  NetLogo 4.04, 66
  Notophthalmus perstriatus, 63
  Notophthalmus viridescens, 63
  pond hydrology, 73–74
  precipitation-based mortality, 80
  randomized initialization, 71
  red spotted newts, 65
  seasonal movement and habitat use, 64
  sensitivity analysis, 80
  state variables and scales
    agents and patches, 67, 68
    behavior trigger, 67
    canopy cover, 69
    catchment area, 69
    depth, 69
    home pond location, 68
    lifestage, 67
    model landscapes, 67, 68
    patch type, 68
    pond area, 69
    rainfall sensitivity, 68
  user-specified initialization, 71–72

U
U.S. Geological Survey, 186

V
Variable time step method, 14